R. A. Buchanan

THE POWER OF THE MACHINE

MACHINE

The Impact of Technology from
1700 to the Present

VIKING

VIKING

Published by the Penguin Group
Penguin Books Ltd, 27 Wrights Lane, London W8 5TZ, England
Penguin Books USA Inc., 375 Hudson Street, New York, New York 10014, USA
Penguin Books Australia Ltd, Ringwood, Victoria, Australia
Penguin Books Canada Ltd, 10 Alcorn Avenue, Toronto, Ontario, Canada M4V 3B2
Penguin Books (NZ) Ltd, 182–190 Wairau Road, Auckland 10, New Zealand

Penguin Books Ltd, Registered Offices: Harmondsworth, Middlesex, England

First published 1992
1 3 5 7 9 10 8 6 4 2

Filmset in 12/15pt Monophoto Sabon by Selwood Systems, Midsomer Norton, Avon
Printed in Great Britain by Butler and Tanner Ltd, Frome, Somerset

A CIP catalogue record for this book is available from the British Library

ISBN–0–670–836567

for Brenda

again

Contents

Contents

List of Plates

List of Figures

Preface

The idea of technological revolution helps to define and explain one of the most important phenomena of modern life. This is the fact that we live, in the twentieth century, in a process of profound and continuing change, which is different from the cyclical change of the seasons or the natural processes of ageing. It is, rather, a change in the conditions of life; life has become a process of permanent transmutation, and this makes it increasingly difficult to understand what life was like in times – most of the past experience of humankind – when such change was not endemic. This process of change has been driven by innovations in sources of power, in manufacturing techniques and in the means of transport and communications – in a word, by technology. To be sure, the process of technological change has been going on for a very long time, since the first men and women struggled to take advantage of their environments, and has acquired growing momentum over many millennia. But in the last three centuries it has accelerated in an unprecedented manner, and the consequent changes in social and human conditions have been so far-reaching that the whole process justly deserves to be regarded as one of technological revolution.

Despite the enormous importance of technology in the modern world, any attempt to examine it tends to be obfuscated by the sheer complexity and technicality of the subject. It was not easy in the eighteenth century to understand how a steam engine worked, and however dexterous contemporary secretaries may be in handling their personal computers, few of them have much comprehension of how they operate. The properties of electricity, nuclear energy and complicated chemical reactions have been brought under human control, but even scientists familiar with these concepts find it difficult to express them in terms which the lay person can understand. The image of the 'black box' has become a pervasive symbol of technological mythology, representing the arcane wizardry by which the technologically competent are able to get things done for the rest of us, who must rely upon them. The best-known black box is that carried by civil airliners to maintain a running record of instrument readings in a secure and recoverable form so that basic information is available in the event of an accident, but the same mystique applies to a digital wristwatch keeping perfect time with a quartz crystal agitated by a minute battery and with a liquid crystal display of figures.

This sense of secrecy about technology arises from the complexity of the techniques involved, but it is maintained in part by the willingness of many people to regard the whole subject as one of intimidating mystery. It is certainly beyond my competence to reveal the secrets of modern technology, but it is possible nevertheless to do a certain amount of demythologizing, and thus to overcome the inhibitions which so many people feel about coming to terms with technology. In short, it is the object of this book to explain how technology has affected life in the modern world, and thereby to come to an understanding of its potentialities and dangers. We all have to live with technology, and it is better that we should know something about it rather than spend our lives in ignorance or fear of something which is so important for our well-being.

This book has undergone a long gestation, and represents the culmination of several converging lines of development. In the first place there has been my long-standing involvement with the study of the history of technology at the University of Bath and through various national and international organizations, and I am particularly grateful for the stimulus which I have received over the years from fellow members of the British Newcomen Society, the American Society for the History of Technology (SHOT) and the International Committee for the History of Technology (ICOHTEC). These associations have brought me into contact with an ever-widening community of scholars and practitioners who have experienced the fascination of the history of technology, and who have encouraged me to proceed with my own researches in the Centre for the History of Technology which I established in 1964 at the college which was shortly to become the University of Bath.

Then there has been my involvement with industrial archaeology, which has made a material contribution to the shape of this book because it began, in a sense, as an attempt to bring up to date my successful Pelican Book, *Industrial Archaeology in Britain*, first published in 1972, with a second edition in 1982. I was closely involved in the development of industrial archaeology in Britain, becoming the second President of the Association for Industrial Archaeology (AIA), and my previous book reflected this commitment. But although the present study began with the idea of preparing yet another edition of the industrial archaeology book, it soon became apparent to me that it would be more appropriate to take the opportunity to look behind the physical remains of technological developments (the proper concern of industrial archaeology) to examine in a completely new survey the relationships between technology and society.

And thirdly, there was the more humdrum need to prepare classes for a course, 'Technology in the Modern World', which brought into focus for me several big issues of a historical and environmentalist nature and caused me to think through some aspects of technological revolution which I had not done before. For this I am grateful to those students who have worked through the subject with me, and also to the handful of colleagues, research students, Visiting Fellows and Rolt Memorial Fellows who have helped to sustain a base for the study of the history of technology at the University of Bath at a time when other developments have not made it propitious to consider any substantial growth in historical studies.

The experience of teaching the subject to students with little previous knowledge of technology influenced me in two particular decisions about the form of this book. One was to limit the number of references in the text, which are usually only made when a direct citation is appropriate, and to place all my bibliographical notes with these references at the end of the text. The other was to incorporate some graphics which I have found useful as teaching aids. In doing so, I would like to stress that these figures – apart from those based on stated statistical sources – are intended only to give some visual 'shape' to difficult subjects, and should not be regarded as possessing any precise descriptive qualities. As far as the inset of photographs is concerned, I am grateful, as ever, to Colin Wilson, Photographer at the University of Bath, for his valiant work with my prints and transparencies. I also thank all those who have given me permission for the reproduction of photographs, as acknowledged in the list of plates. And I thank Aileen Mowry for painstaking secretarial assistance, and the librarian and staff of the University Library for their careful attention to my requests for help.

Finally, I thank my wife, Brenda Buchanan, for all her help and support in working on the theme of this book over the years. I dedicated my first book, *Technology and Social Progress*, to her in 1965, and I dedicated my book on *Industrial Archaeology in Britain* to our sons, Andrew and Thomas, at a time when my work was very much the product of a family enterprise as far as the choice of holidays and outings was concerned. Now that these young men are well settled into their own chosen careers, my wife and I have found ourselves once again relying on each other for scholarly commentary and appraisal. So it is with deep gratitude that I dedicate this book to her.

R. A. Buchanan
University of Bath
May 1991

Part One

BACKGROUND – DEFINITIONS AND CHRONOLOGY

1. *The Nature of Technology*

Technology is the study of human techniques for making and doing things. Strictly speaking – and the English language is rarely used with such precision on this point – technology is not concerned with the mastery of the techniques which are the subjects of its study: these are likely to require a large amount of specialist knowledge which can only be acquired by a long apprenticeship. Technology deals instead with the when and the where, the how and the why of such techniques: it aims at interpreting them within a social context, at explaining how they arose and flourished, the manner of their interrelationships, their ramifications and the reasons for their decline. Technology is concerned with understanding techniques within the environment of their social development. Skill in the particular techniques concerned, although not undesirable, is rarely accessible and is never essential.

From this it follows that technology is a humanistic study and a social study, because it deals with specific forms of human behaviour in society. And as it deals with changes in techniques over time it is perforce a historical study, so that the term 'history of technology' represents a necessary extension of the investigation of technology. As we will use the term in this study, the history of technology can range over the whole history of humankind, because humankind is that series of species, culminating in *Homo sapiens*, distinguished by the capacity to make things – to make tools and create artefacts. The history of these species over the last two or three million years is thus properly a field of interest to the historian of technology. In this book we will be concentrating on the last three centuries of this development, but it will help to keep the extraordinary technological progress of these centuries in perspective if we recognize at the outset that the study has a dimension of profound antiquity. Technology, that is to say, is not merely a product of modern industrial society. It is as old as human species, so that we should expect to find significant continuities in the history of technology.

While all human techniques are the proper study of technology, common scholarly usage has focused attention on some techniques rather than others. In particular, it has become customary to limit the scope of technology to those techniques concerned with harnessing or generating power, together with the productive and manufacturing techniques, and the constructional, transport and communication techniques to which this power is applied. All other techniques, of a linguistic, artistic, cultural or any other special nature, have acquired their own particular disciplines, and are only of peripheral interest to the historian of technology. This limitation is welcome, for even this narrow conception of technology covers an enormous field. The study of sources of power is the core of the history of technology, but this itself is a subject of many facets, and is one of crucial importance to the understanding of human history. When the application of power to all forms of production, transport and communication is included, together with the need to understand these processes in their social context, it is evident that the history of technology is a subject of substance and significance.

The history of technology is thus a part, and an important part, of the study of history. As such, it is subject to all the rules of historical scholarship, which means that it depends upon the careful posing of questions to be answered by the critical exam-

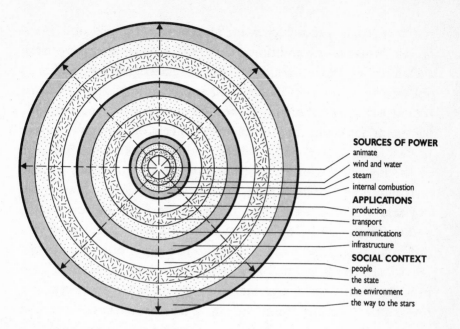

SOURCES OF POWER
animate
wind and water
steam
internal combustion
APPLICATIONS
production
transport
communications
infrastructure
SOCIAL CONTEXT
people
the state
the environment
the way to the stars

Figure 1. Elements of technological revolution

ination of all the available sources of evidence. This evidence may be of various types – documentary, physical, oral and cultural. Most traditional historiography depends heavily upon documentary evidence – the written and printed records left by the participants in a historical period or event, which may be of a personal or official, an informal or a formal nature, but all of which are subject to testing for accuracy, reliability and significance according to well-established principles of historical research. It is true to say, moreover, that all other forms of historical evidence – physical, oral and cultural – must be converted into a documentary form in the shape of a written report or statement before they can be readily used in a historical investigation. But it is worth retaining them as separate categories because they remind the historian to look beyond the documents, narrowly conceived, which come into his or her hands, and to ask whether there is any additional information to be gathered – from the physical evidence of landscapes, settlement patterns, buildings and artefacts; the oral evidence of surviving participants in historical events; or the cultural traditions and practices available for examination by anthropologists and other experts – before undertaking the reconstruction of the past which is the purpose of the historical exercise. This is particularly important for the historian of technology, whose subject-matter consists largely of physical artefacts which have to be understood and explained, and in this task the physical evidence of such artefacts, either directly or at second hand through drawings, photographs or the record of a reporter, will have a special significance. Over long periods of early technological development, indeed, physical evidence is the only sort of evidence at the disposal of the historian, so that its collection and evaluation is indispensable to any historical reconstruction.

The implication of this assertion that physical evidence is important to the historian of technology is that there is a necessary archaeological dimension to the subject, and that archaeology itself is to be properly understood as part of history. History is concerned with the whole of the past, in so far as any sort of evidence is available to allow a reasonable attempt at reconstruction. Archaeology is concerned with the interpretation of physical evidence in the course of a historical reconstruction, and has acquired some subtle and powerful techniques such as stratigraphic excavation, carbon dating, dendrochronology and pollen analysis in order to achieve this objective. It is particularly important for those periods of human history to which the misnomer 'prehistory' has been traditionally applied, for which there is no documentary evidence other than that provided by archaeological reports. It is also important, however, for those aspects of recorded history which have tended to receive least attention from traditional chroniclers, amongst which industrial techniques figure prominently, so that industrial archaeology has come to represent a significant extension of the artefactual study of the history of technology.

Technology, as well as having an intimate relationship with history and archaeology, is also closely linked to science. In current usage it is common to speak of 'science and technology' as if they were virtually synonymous. This, however, is misleading, because they represent different entities. While technology is concerned with making and doing things, science is concerned with the systematic understanding of men and women in their environment. Human beings could begin to make and do things long before they had more than the most partial and primitive explanations of the materials at their disposal, so that technology is much older than science. Science in a recognizable form as an understanding of men and women in their environment achieved by a systematic explanation is only as old as civilization – a mere four or five thousand years – because it was only in civilized communities that the techniques of literacy and numeracy came into existence to permit accurate measurement and recording, thus laying the basis of all scientific examinations. Modern scientists have pursued their objective with an increasingly sophisticated range of concepts and methods, and on a much larger scale than ever before, but the objective itself has remained the same – to penetrate the realities of the cosmos of which humankind is a part. In this exercise technology has played a significant part, by providing the scientist with instruments and methods, and it has been deeply influenced by the successes of science. Nevertheless, it remains qualitatively different from science, so that we are justified in studying the history of technology as something quite distinct from the history of science, even though it will frequently be necessary to comment on aspects of their relationship, which has grown ever closer in the last three centuries.

Before embarking on an examination of technology in the modern world, however, let us go back to the beginning, in order to set the time-scale and chronology of our subject. Human beings acquired their first techniques of tool-making and the control of fire over many millennia in the course of the last two or three million years. We know virtually nothing about the circumstances of these inventions, nor of the associated development of articulate language, but their significance was momentous because they made it possible for men and women to assert an increasing impact on, and eventually a control over, their environment. These early techniques were acquired by hominids ranging in small groups over the most congenial parts of subtropical Africa: they were probably lost and rediscovered several times before the numbers of human beings increased sufficiently to ensure lasting continuity of know-how – skills being passed on in family groups constituting an incipient form of apprenticeship. Unlike other species which possess a high degree of instinctual skill, or even like those higher primates which are capable of forming a simple tool when an occasion requires it, human beings have had to learn their skills, and this has implied a facility for passing them on from one generation to the next. Slowly, therefore, but none the less surely, people acquired the tools which enabled them to impose themselves on the world around them.

The result of this process was that hominids survived and prospered. Although weak and vulnerable, subsisting on the margin of the subtropical forest and grassland, and a prey to the big cats and other natural hazards, they created the wherewithal to defend themselves and to increase their food-gathering potential. So humankind increased in numbers and spread out into more challenging environments, where the invention of new tools made

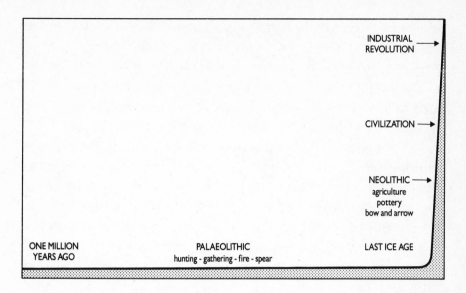

Figure 2. Technological development over one million years

possible the beginning of agriculture through clearing, ploughing and draining the soil, and reaping and processing the crops which they grew. This transformation from a nomadic existence in search of food to settled communities of farmers and stockbreeders has been described by archaeologists as the 'Neolithic Revolution', because it marked the latest (and therefore the newest) stage of the Stone Age, and because it represented such a tremendous change in patterns of life that it deserves to be compared with other social revolutions. It occurred first alongside the great rivers of the middle latitudes – the Nile, the Tigris and Euphrates, the Indus, the Ganges and the Yellow River – where there was fertile soil which received regular enrichment from the flood waters of the river, and which was not covered by dense forest.

Such circumstances were particularly congenial to the primitive techniques of early agriculturalists, so that it was here, over several

9

millennia from about 10,000 BC, that the Neolithic Revolution began, spreading out gradually to most of the Euroasian and African land mass thereafter. The transformation to settled agrarian communities was accompanied by an acceleration in the growth of technical skills. The working of stone for tools and weapons was extended and refined, and the possession of domesticated animals promoted the skills of converting animal fleeces into fibre for textiles. Improvements in the control of fire led to the creation of kilns and furnaces for the manufacture of bricks and ceramics, and eventually for the working of metals. With the acquisition of metalworking techniques, recovering the metals from their natural ores and then forming them into tools and other desirable artefacts, human societies were poised for another profound transformation to the beginning of urban communities.

Perhaps the most far-reaching consequence of the Neolithic Revolution was the expansion of the human population. This was achieved through the establishment of safer conditions of life, with regular supplies of food and protection against predators, including rival human communities. Such conditions required social order, to maintain an equitable distribution of food and to provide defence. This stimulated the emergence of specialized skills, so that warriors could protect the community and craftsmen could concentrate on their skills, freed from the need to grow food, which remained the basic guarantee of the existence of the community. Whereas most of the early skills of the Neolithic Revolution such as the making of textiles and ceramics, and the crafts of baking, brewing and distilling, could be practised within the agricultural community, and were usually performed by the women, the introduction of metal crafts promoted new social relationships. Metalworking involved hard physical labour and metal goods were highly prized, so the skills involved tended to be practised by men, and to acquire a distinct social esteem. The search for sources of metals, moreover, involved exploration and trade, because they were rarely to be found in sufficient abundance precisely where they were most required. Some metals, such as gold and silver, always enjoyed a scarcity value, but some more generally useful metals like tin, essential for the manufacture of bronze (an alloy of copper and tin), were difficult to find around the eastern Mediterranean, and had to be sought at a distance.

The Bronze Age civilizations of the second millennium BC were prepared to sustain elaborate trading expeditions around the Mediterranean and even beyond, as far as the Cornish peninsula in Britain, to find the tin which was an ingredient vital for their weapons, tools and ornaments. The social organization of an enterprise on this scale, with its ships, merchants and cash economy, was characteristic of the urbanized, civilized communities which had emerged from the agricultural societies of the Neolithic Revolution in response to the new metal-using economies. Archaeological chronologists have therefore chosen to distinguish these more advanced societies by the metals which they worked: the 'Copper Age', the 'Bronze Age' and the 'Iron Age'. On this reckoning, the Copper Age coincided with the appearance of the first urban civilizations in Egypt, Sumeria, the Indus valley and parts of northern China close to the Yellow River, which occurred during the fourth and third millennia BC. The metalworking skills were simple, being concerned with the reduction of comparatively soft metals such as copper and gold, and the subsequent making of metal artefacts by moulding, hammering, rolling and drawing out. As techniques improved and the properties of other metals were investigated, the same civilizations moved into the Bronze Age in the second and first millennia BC. Being harder than copper, bronze provided a useful material for cutting tools and weapons, and ingenious means were devised for casting it in molten form from more efficient furnaces than any available hitherto. Possession of metal goods was not generally diffused through these early civilizations, and was probably a sign of social distinction. In contrast, iron goods came to be widely held in the Iron Age, which has caused some commentators to regard it as the 'democratic' metal. Although this is an over-colourful expression, it serves to draw our attention to the social ramifications of technological change.

The Iron Age appears to have begun outside the earlier centres of civilization in what is now Asia Minor round about 1000 BC and was distinguished by the invention of techniques for working iron ore. This material occurs very widely, being found in an easily extractable form in marshy ground as 'bog iron' as well as in hard-rock deposits. The problem was that an exceptionally high temperature had to be sustained in a furnace for several hours in order to reduce the ore to a workable body of metallic iron. The inauguration of the Iron Age thus depended upon the success of the early iron-workers in constructing efficient furnaces, often on exposed hilltops in order to utilize natural wind currents, or using hand- or foot-powered bellows to give an artificial blast. Once these techniques had been mastered, iron goods rapidly replaced bronze in tools and weapons, for the excellent reason that they took a better edge than bronze, and so they conferred a decided advantage both in clearing forests and in battle. They spread quickly throughout the Middle East and northwards and westwards into Europe. The legendary heroes of Homer's epics, around 1200 BC, were Bronze Age warriors, but the civilizations of classical Greece and Rome, from the fifth century BC, belonged firmly to the Iron Age. Unlike the earlier metals, iron was rapidly adopted for a variety of everyday uses, such as in cooking pots and pans, and in the form of nails for all sorts of building. Iron, in short, affected the quality of life in a way that previous metals had failed to do.

Nevertheless, it is unrealistic to regard iron as being conducive to the growth of democracy in any modern sense, because it had remarkably little effect in transforming political relationships within the classical civilizations of Greece and Rome. These relationships had been determined by the distance of the new societies from Egypt and Persia, by the derivative nature of many of their cultural traditions, and by the lack of political cohesion amongst the city-states which, encouraged by thriving trading links, flourished on the periphery of the monolithic civilizations of the Middle East. Also, in common with other ancient civilizations, the city-states of Greece and the empire of Rome accepted the institution of slavery as a necessary part of the social substructure, providing a labour force for the most menial and physically laborious work. This is not easily reconciled with the view that the use of iron promoted democracy, even though, over a much longer period, the spread of iron-working probably contributed towards a wider social appreciation of the benefits of technological improvements. More to the point, in the context of classical civilizations, is the argument that the institution of slavery acted as a brake on technological innovation. This stresses the fact that the existence of a large pool of slave labour made labour-saving devices redundant, so that there was little incentive to put substantial resources into the development of, say, water power, even though the Romans were familiar with all the basic techniques. Moreover, it can be argued that the possession of slaves imposes an obligation to find work to keep them busy, if only to keep them out of mischief, and thus provides a positive disincentive to developing machines which could make routine labour easier.

It is impossible to prove or disprove this argument, but it seems likely, to say the least, that the heavy dependence upon slave labour in the ancient civilizations retarded technological development, because it is a striking fact that most of the intellectual resourcefulness of these societies was devoted to non-technical objectives of a political, religious and cultural nature. In so far as technology could be exploited to improve fighting power or to construct more impressive buildings, there were significant advances in the ancient world, but in the more routine aspects of labour-saving devices and manufacturing productivity very little progress was made. However, it should be remembered that we are now dealing in much shorter time spans, centuries rather than millennia, so that the overall pace of technological advance was manifestly greater than it had been during the Neolithic Revolution or the Old Stone Age which preceded it.

In discussing the development of technology from the earliest human societies to the Iron Age we have slipped easily into using the concept 'civilization'. It is worth pausing to make sure we understand what we are talking about. A civilization is a large and complex social organization, effective over an extensive territory and over a long period of time, and sharing many cultural traditions. It is distinguished from more primitive societies by the sophistication of its institutions, involving a high degree of functional specialization amongst its members; by a distinct system of social stratification; and by the mastery of literacy and numeracy, giving rise to the creation of historical records and the emergence of systematic scientific activity. All civilizations are built upon a well-developed system of wealth creation which can ensure the production and distribution of food and other essentials of life, and this normally involves trading relations and urbanization. Technology plays a vital part at this level, in the shape of agricultural techniques, constructional skills in the provision of buildings, irrigation works and aqueducts, and in the improvement of roads and ships.

Most human societies since the fourth millennium BC have been part of one or other of the civilizations which have flourished since

14

then. Historians have been able to study the rise and decline of
many such civilizations, and occasionally the catastrophic collapse
of a civilization has provided an opportunity for a significant
regrouping of human resources. Such a collapse occurred when
the Western Roman Empire was overwhelmed by the invading
barbarians in the fifth century of the Christian era. Out of the
ensuing chaos in Western Europe there eventually emerged a
distinctive new civilization. This is the Western Civilization which
has taken the lead in world technological development over the
last millennium, and it is with the most recent stages of this process
that we will be mainly concerned in this book. But it is worth
sketching in briefly the main lines of development, in order to
explain the nature of the roots from which modern technology
has sprung.

Western Civilization was still, in the conventional archaeological chronology which we have adopted, part of the Iron Age
when it emerged as an identifiable entity in the form of medieval
Christendom. The societies which fused together to form relatively
stable political units after the confusion into which much of
Europe lapsed with the withdrawal of the Roman legions and the
settlement of barbarian tribes from beyond the Danube and the
Rhine practised techniques of iron-working which had not
developed any outstanding refinements over the past two millennia. Important new uses were found for iron, such as in the
provision of horseshoes, but the methods of reducing iron ore and
of forging iron tools and weapons underwent little change. In
particular, all the iron available to the Iron Age before the end of
the Middle Ages was in the form which we know as wrought iron:
no way was known of producing cast iron in bulk. In the fifteenth
and sixteenth centuries, however, a very significant new technique
appeared, which greatly increased the versatility of iron. This was
the 'high furnace', or 'blast furnace', a much larger structure

15

than the traditional 'bloomery furnace', with a constant draught provided by water power to raise the temperature above the melting point for iron, which meant that it became possible to produce iron in molten form and to cast it into a wide variety of shapes, including cannon for heavy artillery and columns for building work.

This became known as the 'indirect method' of iron production because, in order to obtain the traditional 'malleable' or 'wrought' iron, which was capable of being worked by rolling and hammering it into the shapes required, or refined further to make it into steel, the cast iron from the blast furnaces had to be laboriously reprocessed. Nevertheless, the demand for cast iron was such that the new process rapidly superseded the traditional 'direct' method, and the use of iron underwent a corresponding expansion. So marked was the change from one method to the other that, by analogy with the 'Old' and 'New' Stone Ages, it is reasonable to represent the change as a shift from the technology of an 'Old Iron Age' to that of a 'New Iron Age', and to interpret the position of our own society, for all its many sophistications in the use of metals and other materials, as being essentially part of this New Iron Age.

While Western Civilization remained – and has continued to remain – one based upon iron-working techniques, it has been remarkable for the proliferation of technical innovations in many other aspects of life. Agricultural productivity, for example, has been transformed by the introduction of new machines and techniques. These go back to the improved systems of crop rotation and the introduction of horses as the major source of animate power in the Middle Ages, and continue through new crops and methods of animal husbandry to the advanced factory farming of modern agriculture. Similarly, techniques of power generation, transport and manufacturing industry have been in a condition of continuous revolution for several centuries. It is against this background of responsiveness to innovation that the emergence of our contemporary technological society has to be seen, and it is not easy to account for the phenomenon. True, Western Civilization has developed without the substructure of slavery which characterized the ancient civilizations (except as an instrument of colonial exploitation), so that there was frequently a shortage of labour which created a demand for labour-saving devices. But a more important factor was probably that Western Civilization lacked the homogeneity of most ancient civilizations; it comprised a motley collection of small states developing in conditions of sharp rivalry for political power, for trade and, after the European Reformation of the sixteenth century, in religion also. This rivalry acted as a stimulus to innovation, and those states which provided greater social mobility or other potential rewards to the successful innovator came to enjoy a distinct advantage over those which were more closed to such influences.

The most dramatic contrast with Western Civilization is shown by its great contemporary world culture, Chinese Civilization. This, of course, is much older than the civilization which emerged on the western seaboard of Europe about a millennium ago: the civilization of China subsumes a whole series of developments in a single area, stretching over at least three millennia, so that it would be reasonable to regard it as a set of different civilizations if it were not for the striking cultural uniformity maintained over this period. One aspect of this continuity was the development of Chinese technology, which far exceeded anything achieved in Western cultures until the end of the European Middle Ages. Chinese metalworkers mastered the technique of raising furnace temperatures sufficiently to produce cast iron long before this had been done in the West, but without any dramatic social repercussions, so that the distinction between an 'Old' and a 'New'

Iron Age has no significance in China. Similarly, the development of printing, paper, gunpowder, the compass and other important innovations occurred in China before they did in the West, but without social disruption.

The best explanation of this evolutionary pattern is probably that the whole of Chinese society was dominated throughout three millennia by the mandarins – a non-hereditary intellectual aristocracy, subject to continuous recruitment by competitive examination, and wielding enormous power and authority in the name of the emperor whom it existed to serve. For our purposes, the significant thing about this ruling mandarinate was its stability: although it controlled all aspects of life in China, it was not hostile to the appearance of occasional technical innovations, provided that they performed a purpose which was useful to the mandarins and could be easily assimilated into the existing social order. Thus Chinese technology developed slowly but steadily, with each innovation being absorbed relatively painlessly into the fabric of society. Ultimately, however, and particularly in contrast to the rapid growth of Western Civilization, this process came to appear as one of stagnation, and Chinese Civilization was thrown on to the defensive by the assertive expansion of the West. Modern China, like the rest of the non-Western world, has been obliged to undergo 'Westernization' in order to seek the cornucopia of material wealth which appears to be the outstanding material achievement of the West. And by doing so it pays tribute to the technological dynamism which has been the characteristic of Western Civilization.

Most other civilizations have been comparatively unresponsive to technological innovation. The extensive culture of world Islam, for instance, played a very significant part in relaying the techniques of the ancient world and of contemporary China to the West, and in some aspects of mathematics, shipbuilding and navigation it was responsible for important innovations on its own account. But like China, Islamic Civilization was unable to keep pace with the explosive expansion of Western technology, and adopted many aspects of Westernized society in order to participate in the material benefits which flowed from mastery of the dynamic technology of the West. Much the same could be said of those Indian civilizations which, inspired by Hinduism or Buddhism, developed rich cultural traditions with a philosophical profundity which was often lacking in the West, but which placed little reliance on technical performance. In terms of increasing their wealth-producing capacity, the modern descendants of all these great civilizations have found it necessary to turn to Western Civilization for guidance and inspiration.

Alone amongst terrestrial cultures, therefore, Western Civilization has achieved power over its neighbours and over its environment by the strenuous application of a dynamic technology which has been encouraged to develop by the social conditions within its member states. Alone amongst world civilizations, it has been dominated by a process of continuous technological revolution, causing a ferment of social disruption and acute internal rivalry. This has not been a comfortable period in world history. On the contrary, it has been fraught by intense rivalry between competing states and this has erupted into a series of violent wars; it has experienced unprecedented pressures from increasing population in conditions of expanding social mobility; and it has undergone profound changes of culture, orientation and aspirations. Nevertheless, it is a period which calls for historical explanation because it has so many unique features, and as it is the period which has shaped contemporary society, the historical explanation is necessary in order to assess the future prospects of our society. Study of the technological revolution which has dominated Western Civilization since 1700 can thus contribute to an understanding of the extraordinary situation, full of dangers and potentialities, in which citizens of Planet Earth find themselves at the end of the twentieth century.

2. The Process of Technological Revolution

The condition of continuing technological revolution which has characterized Western Civilization for the past three hundred years is the product of several converging developments. Some of these have been anticipated in the first chapter of this book, in the course of describing the features of technological growth over many millennia up to the eighteenth century. Amongst the more important factors have been the invention of systematic agriculture, making possible a settled society and creating the potential for an expansion of population; the emergence of large 'civilized' societies possessing specialized social distinctions and the arts of literacy and numeracy; the mastery of ever more sophisticated skills of metalworking and the use of metal artefacts; the development of a spirit of scientific inquiry devoted to discovering more about the nature of human beings in their environment; and, with the rise of Western Civilization, the intense rivalry between conflicting political and religious factions creating scope for divergent views to flourish, and stimulate trade, commerce and colonial expansion. In this chapter we will look more closely at this process of technological revolution, attempting to understand the nature of invention and its diffusion, and reviewing the way in which these factors have operated in our society over the last three centuries.

Some historians have tried to distinguish between invention, representing the origin of a new artefact, and innovation and development, with 'innovation' describing the conversion of an invention into a viable commercial proposition and 'development' covering subsequent stages of improvement in a successful artefact. Other historians have pointed to the asymmetry of technological history in so far as it tends to be written entirely from the perspective of successful inventions, and have argued that it is necessary to give more attention to failed inventions. Both these points are valid and need to be recognized. The distinction between invention, innovation and development is one which is normally convenient, and will be adopted here. As far as the question of symmetry is concerned, we will inevitably be mainly concerned with successful rather than with failed inventions, because it is the successful artefacts which have been responsible for the transformation of the modern world. In contrast with this, failed inventions, however interesting and suggestive as ideas, are historically irrelevant. But we will have cause to consider some of them as stimuli to more successful ideas, or as alternative possibilities to developments which actually occurred.

Taking the process of invention through to successful development as that which provides the primary engine of change in modern society, it is possible to discern certain patterns in its history. It is tempting, indeed, to look for a theory of invention capable of explaining the mechanisms of the process and even of performing some predictive functions; this temptation must be resisted, however, because any theory of this sort tends to distort the extraordinary diversity of human inventiveness by simplifying it and to abuse historical analysis by using it to forecast future developments. It is enough for our purposes to distinguish the factors which recur consistently and regularly in the course of invention and innovation. For one thing, it has frequently been observed that invention responds to the stimulus of social need: in the words of the adage, 'necessity is the mother of invention'.

But when we regard the many pressing social needs in the world today to which there has been no successful inventive response, it is clear that 'necessity' is largely a matter of perception; a need must be felt as such, and it must also be thought to be capable of amelioration by some sort of application. A society which is capable of such application will then generate inventions in order to meet the specified need. Whereas an Eastern peasant, labouring in his paddy-field, is unlikely to articulate a social need in a situation which he has accepted as inevitable, a Western manu-facturer, enjoying a much more comfortable standard of living, will apply himself vigorously to solving a mechanical or organ-izational problem which limits his productivity.

The difference between Eastern and Western attitudes in this caricature is not just one of social conditioning, important although this undoubtedly is. It is also one of social resources. The European or North American knows that, in any situation in which there is a demand for improvement or change, he can call upon a range of economic and social resources. The 'factors of production' of traditional economics – capital, labour and land – are available for his use, as is the social factor represented by skill, passed on by education and training. Many innovative ideas have failed to materialize when these resources have not been available. One of the most inventive minds of Western Civilization was that of Leonardo da Vinci, from whose notebooks it is apparent that he conceived notions for flying machines and submarines, but in the conditions of sixteenth-century Italy the capital, the materials and the necessary skills were simply not available to him, and so his ideas came to nothing. Two centuries later, the Scottish craftsman James Watt might have had a similar experience with his brilliant ideas for a new type of steam engine, but he had the great good fortune to enter into partnership with a Birmingham industrialist, Matthew Boulton, who could supply the capital, the machinery and the skilled craftsmen required to convert Watt's ideas into successful artefacts. No invention can be generally accepted until such economic and social resources have been made available for it.

If social need and social resources may be regarded as pre-requisites for a successful invention, their presence alone will not generate it. Every invention must come from an idea, which means that an inventor must have an inspiration, and we are bound to be less precise in defining the circumstances of such inspiration than those of the more material prerequisites of invention. Thomas Alva Edison undoubtedly had a point in regarding his inventions as the product of 99 per cent perspiration and 1 per cent inspiration, but without the spark of inspiration there is no guarantee that any amount of perspiration will produce the desired result.[1] Edison possessed this inventive genius, and every successful inventor has had a share of it. Whether or not James Watt ever watched water boiling in a kettle and thought of steam power we cannot be certain, but he did describe how the idea for the separate condenser, which revolutionized the clumsy old atmospheric steam engine, came to him in a flash as he strolled on the Green of Glasgow University one sabbath afternoon. Likewise, Wilbur and Orville Wright watched the movements of seagulls and thereby conceived the principles of controlled flight. Many such apparently simple observations – except that nobody had the initiative to act on them before – have led to momentous inventions.

It seems likely that inventions come to the prepared mind. James Brindley is reputed to have wrestled with the novel problems of canal construction by taking to his bed and waking up with the solutions, and there are many testimonies to such enlightenment coming to a well-prepared mind struggling with an unfamiliar problem. Of course, when the problem is unfamiliar it is difficult to know exactly what preparation will be most conducive to its solution, and there are examples of cross-disciplinary invention, when the inventor has brought a new expertise to bear on a recalcitrant problem. Modern industrial enterprises have sought to cater for this contingency by creating research teams of mixed skills to work on the exploitation of new ideas, and there can be no doubt that such methods have reaped very handsome rewards in electronics, pharmaceutics and chemical engineering, amongst other fields. In the last resort, however, invention remains an intensely personal phenomenon: inspiration occurs to a particular inventor, in a particular state of mind and with a special degree of preparation, but the vital human ingredient, the stroke of genius, is essentially unpredictable.

Invention, then, cannot be reduced to a set of neat theoretical categories. Because it relies on the inspiration of human inventors, invention has much in common with art, and is equally unpredictable. This affinity is worth stressing, because in our mechanistic age it is frequently overlooked. An imaginative innovation in the design of a motor car or the structure of a large bridge or the properties of a new plastic has aesthetic qualities representing a combination of creativity with functional elegance: the engineer who designed them must be satisfied that they 'feel' right, even though he is probably unable to explain his intuitive sense. Such qualities cannot be summoned to order, but they can be encouraged, and perhaps the most important of all the prerequisites for a successful invention is the existence of a social *milieu* which is sympathetic towards it. It has been well said that one of the most significant features of modern Western Civilization has been the 'invention of invention', the establishment of a social system which has been supportive of inventors and their ideas. This involves the creation of a system of legal protection through an adequate network of patent regulations, so that the inventor can enjoy the financial rewards from his ideas, and a society which is sufficiently open to allow him other rewards in terms of upward social mobility and public esteem. It also involves active encouragement from financial agencies and, when very large schemes are being undertaken, a degree of state patronage. Thus, even though no 'theory of invention' can properly be defined, it is clear that there is a pattern of prerequisites for successful invention which makes it possible to recognize a package of circumstances without which invention cannot be expected to flourish.

If the nature of invention may be taken as constituting the first great problem of the history of technology, the second such problem is that of the diffusion of innovation. It is a striking fact of human societies that successful inventions spread from their point of origin and are adopted in other places, usually close at hand to begin with, but eventually far removed from their place of origin. The mechanisms of this process of transmission are sometimes obscure, and they may then be difficult to distinguish from invention occurring spontaneously. There was, for instance, a strong tradition amongst European archaeologists that many of the great inventions of classical times were derived from the Middle Eastern civilizations, from which they were gradually disseminated. When such an extraordinary monument as Stonehenge is considered, for instance, it can be argued that the mathematical and architectural skills involved in its construction were too sophisticated to have been indigenous to the illiterate tribal community which, it is assumed, occupied southern Britain in the second millennium BC when it was built. It is supposed, therefore, by those who adopt this argument, that Stonehenge was built under the guidance of master craftsmen who had brought the necessary skills from Eastern Mediterranean civilizations, where they certainly existed at this time. The alternative is to believe that the society of northern Europe, in the centuries before the Roman Empire established its authority over these parts, was less primitive than has been assumed and that the skills emerged spontaneously, only to be lost again after Stonehenge had been built. Both theories require a keen imagination, and it is probably best to suspend judgement until more archaeological information becomes available.

Another example of the diffusionist *v.* spontaneous-generation controversy is the attempt of the Norwegian explorer and anthropologist Thor Heyerdahl to demonstrate diffusionist tendencies on a global scale. He argued that it was technically possible for Polynesian people to have crossed the southern Pacific on rafts using only the ocean currents and primitive sail power and for Egyptian-taught craftsmen to have crossed the north Atlantic on similar rafts; he proved his point by making both crossings himself, so that it must be conceded that Polynesian culture could have crossed the Pacific from South America to New Zealand, and that the building skills of the Maya, Aztec and Inca civilizations could have been acquired from ancient Egypt. The possible transmission of technology from Egypt to America is particularly interesting from our point of view, but it has to be said that as yet no conclusive proof has emerged that such a transmission occurred, and that it is strange that no literary evidence has been found amongst the elaborate hieroglyphic scripts of New World societies if there was any Egyptian tutelage involved. Once again, it seems best to suspend judgement, in the hope that more evidence will become available.[2]

The question of the spread of Chinese inventions to the West is particularly sensitive because it implies a derivative quality about Western Civilization which Western historians have generally been reluctant to accept. We have already observed, however, that Chinese Civilization was one of great antiquity, having undergone relatively continuous and homogeneous development from the second millennium BC, and that it had acquired, under the supervision of the non-hereditary aristocracy of the mandarins, a facility for adopting useful inventions and putting them to constructive social applications. Amongst other techniques, the Chinese had mastered the use of cast iron, the manufacture of gunpowder and porcelain, the construction of mechanical clocks and of windmills, and printing with movable type, before they had been invented in Western Civilization. In view of the undoubted precedence of China in so many areas of technological competence, it is at least possible that they were transmitted to the West, and, although the evidence is rarely conclusive, there is in fact a very strong inference that some such transmission occurred in most cases. In that of gunpowder, for instance, it seems virtually certain that the European inventors were inspired to experiment with formulae brought back from China by returning travellers. The cases of the windmill and printing are not so clearly defined, and in clock mechanisms the Europeans developed quite different control devices from those employed in China, but as far as porcelain was concerned the West made no secret of its wish to emulate Chinese methods. What is now beyond reasonable historical doubt is the fact that Europe was powerfully stimulated by Chinese technological superiority during the Middle Ages, and that this influence remained strong thereafter in ceramics and some other areas, even though the West had acquired its own momentum in technological innovation after AD 1500.

Over against the theories involving transmission of inventions from one or two dominant sources, there is an alternative interpretation which places its main emphasis on the universality of inventions. According to this view, inventions tend to occur naturally and spontaneously once the conditions are appropriate. It is important to recognize the strength of this argument, because

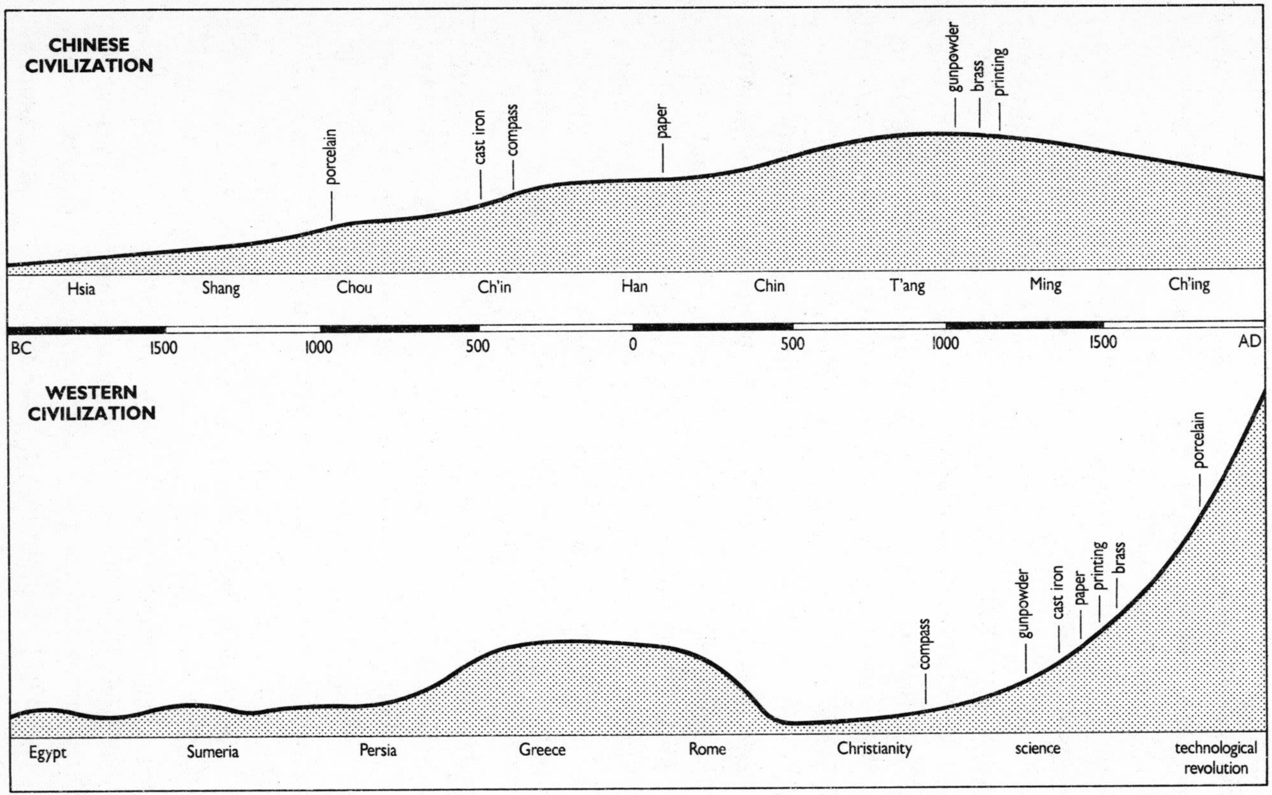

Figure 3. The comparative development of world civilizations (vertical axis represents stability, complexity, urban development, etc.)

it is based upon characteristics of inquisitiveness and adaptability which, as we have seen, have been ever-present features of man-like species and, outstandingly, of *Homo sapiens*. It is likely, moreover, that men and women have frequently had ideas for inventions, and they have subsequently been lost or forgotten, so that there has been a large amount of reinvention over time. However, the development of civilized societies tended to insti-tutionalize technologies through literacy, numeracy and edu-cation, so that inventions were less likely to be lost and became more widely available for imitation, which meant that the mech-anisms of transmission could operate on all inventions, whether they were newly devised or whether they had been derived from some older tradition. While acknowledging the possibility of spon-taneous invention occurring at any time or place in human society, therefore, it is necessary to take account of the fact that, in civilized societies at least, inventions are readily transmitted by various means – by what may be called the 'mechanisms of transmission'.

These mechanisms include the migration of craftsmen, the dis-tribution of artefacts, information conveyed by travellers and information contained in journals, patents and other publications. We will have cause to observe all these processes at work, but it is useful to distinguish them as factors in technological revolution. The ability of craftsmen to transfer their skills from one environ-ment to another has been a most potent mode of transmission. They can be induced to move by bribery or by the prospect of financial reward; they can be moved by slavery or punitive transportation; or they can move in order to avoid political or religious persecution, or simply from a desire to seek a new life elsewhere. Whatever the reason for the move, they take their skills with them, and while possibly impeded by lack of adequate tools or materials, they are capable of re-creating their technological expertise in the new environment. The case of Samuel Slater and others who carried their skills as operators of the new textile machines across the Atlantic at the end of the eighteenth century, and thereby accelerated the development of the textile manufacturing industries in the United States, has been well documented.[3]

It is also possible for artefacts to be the inspiration for the transmission of a technology. When Western mariners first saw a Chinese compass, they took pains to procure and to replicate this invaluable aid to navigation. Throughout human history, the experience of an impressive new artefact has stimulated attempts to copy it, and even though the copies will vary in quality, some of them may actually improve upon the original and so represent a different or better quality of product. Sometimes even a model of the full-scale artefact is sufficient to inform an intelligent craftsman of what he needs to do to make a copy; until quite recently, models of ships have performed this sort of function in the shipbuilding industry. Then again, the return of travellers from foreign parts has been an important source of information for the transfer of technology. The role of travellers such as Marco Polo, who visited China in the thirteenth century, was significant in providing knowledge about the wonders of oriental technology, and travellers of all sorts – diplomatic, commercial, tourists – have moved between the nations of the West and provided an invaluable source of technological know-how, sometimes illicit, but none the less serviceable for that. Published sources, too, in the shape of official documents such as patent specifications, professional papers such as trade journals and periodicals, and the myriad other forms of printed information, have also acted as an effective conduit for the transmission of technology.

All these modes of technological transmission are social devices: they exist because technology occurs within a well-defined social framework which encourages, or at least does not prevent, an easy exchange of personnel, artefacts and information. This matrix of social relationships indicates a third major area of concern in the history of technology: after the problems of invention and the transmission of technological knowledge, we have the social ramifications of technology. There can be no doubt that technological innovation has been a powerful agent of social change, even though the nature of this relationship is sometimes extremely complex, making it difficult to disentangle causal linkages. Technological innovation disturbs 'natural' population mechanisms, for example, by enabling more food to be produced from the same acreage and by removing causes of disease. This in turn stimulates urbanization and disrupts traditional patterns of life and the organization of labour. Similarly, advances in military technology alter political balances of power and thus encourage rivalry and war, while improvements in transport and communications break down barriers of space and time. All these developments promote social change, often in ways which are unexpected and unintended, and make a direct impact on value systems of traditional religions and on artistic and literary imagery. Historians of technology have a responsibility to explore these ramifications of their subject, even if they cannot be expected to produce answers to all the problems involved.

There have been several attempts to impute a systematic pattern to the processes of social change derived from technological innovation. One of the most elegant of these was that expounded by Lewis Mumford in his seminal work *Technics and Civilization*, first published in 1934. Using a wealth of illustrative material, Mumford distinguished a series of cultural stages in the development of Western Civilization, related to the technologies on which they flourished. These were, first, the 'Eotechnic' phase, based upon wood and water power, which persisted until the eighteenth century, and in which the technologies were still essentially simple;

this was followed by the 'Palaeotechnic' phase, based upon the exploitation of coal and iron, in which Britain took the lead until the later decades of the nineteenth century; and then the 'Neotechnic' phase when, with the application of electricity and the internal combustion engine, the lead passed to other European nations and to America. Finally, he speculated about the emergence of a 'Biotechnic' phase, bringing a more harmonious relationship between man and technology than had existed before. Although full of brilliant perceptions and eminently suitable for pedagogic presentation, this is primarily a descriptive pattern giving a vivid account of different phases within the evolution of modern society, rather than an explanation of why this evolution occurred: the explanatory parts are the least satisfactory, tending to be rhetorical and leaving important questions unanswered.

Several other commentators, both before and after Mumford, have attempted to provide systematic accounts of technological revolution. Outstanding amongst the earlier systems was that of Marxist analysis. Marx and Engels were fulsome in their recognition of a technological component in the process of social change, particularly on the role of mechanization in transforming patterns of work from domestic to factory organization, but their emphasis was always on the political revolution promoted by such developments, and the effectiveness of their arguments was largely vitiated by their insistence on the polarization of participants into conflicting social classes, which oversimplified the relationship and minimized the direct participation of individuals in the process. Later writers in this tradition have tried to establish a correlation between the frequency of inventions and the rate of socio-political change, but the results of such exercises remain problematic.[4]

Most recent discussion of technological systems in relation to society has come from social scientists – from economists keen to apply mathematical techniques of measurement and analysis, and from sociologists anxious to demonstrate the 'social construction' of technological change. Both have provided some useful ideas, but neither has produced a satisfactory explanatory system for the process of technological revolution. Economic analysis has been most valuable in stressing the importance of accurate quantification as a basis for theoretical speculation. But its application of counterfactual theory (i.e. attempting to determine what would have happened if a particular innovation had *not* been introduced) to the explanation of problems in technological history such as the supersession of water power by steam power and of canal transport by rail transport has been less helpful, because the calculations can never be sufficiently refined to encompass the infinite number of variables involved.[5] As far as the contribution of the exponents of the 'social construction' of technology is concerned, this has generated considerable enthusiasm and warmth, but as yet remarkably little light. The trouble here has been that social scientists have tended to approach problems in the history of technology with an extensive apparatus of theoretical categories and 'empty conceptual boxes' which have then been filled from the available historical material. This is hard on any evidence which does not fit into the preconceived boxes, and too often the 'solution' to a historical problem is assumed in the structure of the theory by which it is investigated. While it is certain that every historical investigation requires some measure of preliminary theoretical organization, this should never be permitted to dominate the inquiry, and in this respect the endeavours of the social scientists require refinement.[6]

The concept of technological revolution is itself a theoretical construct and must be treated with some care; amongst recent accounts of modern technological history of a comparatively orthodox narrative nature there has been a tendency to avoid it.

James Burke, in his very successful television series *Connections*,[7] made much of the way in which an invention in one field could produce quite unexpected developments in other fields not necessarily directly related to it. He began his review with a graphic account of the great power failure in the north-eastern states of the United States in 1965, when civilized life virtually came to a sudden end, because the loss of electricity brought black-out, stalled elevators and traffic mayhem to the largest metropolitan concentration in the world. He traced the cause of this catastrophe through a chain of interconnected circumstances to the failure of a single contact-breaker. This set the pattern for a selection of such linkages, ranging across the whole history of technology. The result was entertaining and, to a degree, informative, but in the end the excitement palled because the viewer or reader came to feel that it was possible to link any effect with any cause, depending only on the ingenuity of Burke himself. As for the history of technology, it came to seem a mishmash of contrived teleological relationships which lacked any general organizing principle.

More recently, Arnold Pacey has argued a similar theme in his book *Technology in World Civilization*,[8] but he has done so with more scholarly thoroughness than Burke and has produced a coherent interpretation of the history of technology. The main strength of the work is its demonstration of continuing interchanges in technological development between the main cultural traditions of the world. In pursuing this idea of 'dialogue or dialectic in technology' he is particularly good at showing how innovations passed between Asian, Chinese, Islamic, Indian, African, American and, of course, Western cultures. It is an impressive performance, but it does involve a certain blandness: technological innovation is reduced to explicable ordinariness, with little sense of the extraordinary revolutionary quality of so much modern technology. This, however, is just the point at issue: while Burke and Pacey both stress the interconnections of technological history, they do so by abandoning the peculiarities of modern Western technology, whereas the notion of technological revolution is concerned to focus attention on these as the most significant aspects of the narrative.

Revolution can be seen as a sort of accelerated evolution, and the view of the history of technology as an evolutionary process has been strongly argued by George Basalla in his recent work, *The Evolution of Technology*.[9] 'If technology is to evolve,' he affirms, 'then novelty must appear in the midst of the continuous' (p. viii). Although aware that a close analogy with Darwinian evolution of plant and animal species is impossible, for the intensely human reason that technological developments interact with each other in a way which botanical and biological species never do, he insists that the metaphor is useful in maintaining a substructure of continuity to his historical explanation, rather than one marked by discontinuities and pristine innovations. This objective is acceptable, but it should not be necessary in achieving it to deny that some inventions are more novel than others, and that some innovations have produced a disturbing or dramatic effect while others have been assimilated without trouble. In other words, while it may be argued that technological history in general displays strong characteristics of evolutionary interconnectedness, this does not preclude the possibility that there has been something truly revolutionary about developments in Western Civilization over the past three hundred years.

Technological revolution is a process rather than an event. It is the process of profound and continuing change in the technical infrastructure of human societies whereby men and women have transformed the ways in which they have made and done things, thus increasing their capacity for wealth creation and self-destruction. The process has been going on for many centuries, since at least the development of settled farming communities in the Neolithic period (as illustrated in Figure 2), but in the last three hundred years it has been accelerated in the course of industrialization, and has become identified with the world domination of Western Civilization. This identification is effectively unique to Western Civilization, because even though other cultures have made distinguished contributions to the process of technological revolution, it is only in the West in modern times that it has become the distinctive feature and outstanding quality of a civilization, providing both its major blessings (such as a superior capacity for raising the standard of life) and many curses (including its facility for mass destruction).

Technological revolution has not been the only factor of social transformation in Western Civilization. This is not an attempt to revive a discredited mono-causal explanation for everything that has happened in modern history. Nevertheless, while other factors – town life, a market economy, population pressures, religious motivation and so on, have featured in all civilizations at different times and places, the systematic exploitation of technological innovation on a vast scale is found only in modern Western Civilization. It is this fact that justifies us in focusing particular attention on the phenomenon of technological revolution. For better or worse, the world culture of Western Civilization is being determined to an unprecedented degree by its commitment to technology, and it is the exploration of the causes and consequences of this commitment which is the main theme of this book.

It should be observed that the term 'technological revolution' is used here in the singular, to describe a comprehensive and continuing process. If it is objected that this use lacks the implication of sudden discontinuities of practice associated with some political revolutions, it must be remembered that many of these apparent political discontinuities are less drastic than may be apparent at first sight, and also that the profound changes associated with technological revolution are such that they should be seen to be as revolutionary as any other developments in history. It is not worth great semantic debate, but the view held here is that historical revolutions are characterized less by discontinuities than by novelties and social change, and the process with which we are dealing has certainly plenty of these to offer. In summary, the arguments for technological continuities, evolution and interconnectedness which we have just discussed are all accepted here, but it is maintained that the social changes induced by these developments are of such a profound and coherent character that they deserve to be identified as part of a continuing process of technological revolution.

The coherent character of technological revolution derives from the common experiences of invention, innovation and development, and of transmission and its social implications, shared and mutually encouraged within the framework of Western Civilization. We have already discussed the main features of these aspects of the process of technological revolution, but in concluding this chapter it is worth stressing two elements which have previously only been mentioned in passing: the concepts of the 'ratchet' and the 'package'. The image of the ratchet is firmly mechanical rather than biological, and is intended to convey the impression of a mature innovation providing the gears which engage with novel situations in order to promote new inventions, which may in time displace the mature development, or which may stimulate development in an entirely different field. The history of invention is full of such stimuli – the development of a boring machine for cannon being readily adapted for use in boring cylinders for steam engines, or the development of the internal combustion engine making available a sufficiently light powerpack to be applicable to the first flying machines – both provide good examples of the process. The important aspect of the process from our point of view is that the ratchet effect is made possible by the conditions of advanced technological revolution, in which inventions are regarded as familiar and are readily tested for applications. Sometimes, as Basalla has shown, in instances such as Edison's phonograph in the 1880s or the tape recorder in the 1950s, the best application for promising inventions has not immediately appeared, and it has been the ratchet effect which has encouraged developments which did not occur to the original inventors. Such support for new ideas is lacking in traditional societies which do not have much expectation of benefits to be derived from inventions, and its existence is a strong indication of the presence of technological revolution.

The concept of the package is an extension of this idea, being intended to suggest the importance not only of the mutual inter-action of ideas in the process of invention, innovation and develop-ment, but also the existence of supportive social conditions. To become successful, an invention has to be more than a good technical idea. It requires backers to provide capital, labour and skill; it needs favourable market conditions; and it depends upon the ambience of an encouraging environment in which entre-preneurs can be persuaded to make resources and markets avail-able without fear of political or religious persecution. The existence of such a combination of conditions constitutes a package which is conducive to technological revolution, whereas the lack of it makes any new development difficult, if not imposs-ible. Conditions in contemporary China provide an outstanding example: there is plenty of desire for the material benefits of technological revolution, without as yet much understanding of the nature of the social package which is necessary if they are to be achieved. This is not to say that the package must always be the same, and there is certainly a wide range of conditions which will promote development, particularly when there is an incentive geared to defence and the possibility of war. But in most normal situations, these conditions will imply a significant degree of liberalization in order to provide the necessary encouragement to inventors and entrepreneurs. It is important, therefore, to be aware of this package in any attempt to understand the success or otherwise of inventions.

Let us summarize the argument of this chapter. The dominant assumption of this approach to the history of technology is that we are dealing with a process of technological revolution: that is, with a series of intimately related and continuing changes, promoted by technological innovations and of such a nature that they are fundamental to explaining the developing quality of life in our society. It is further taken to be beyond the bounds of legitimate historical doubt that this process has been going on for some centuries: slowly but steadily since the end of the Middle Ages, and accelerating dramatically since the eighteenth century. Moreover, it is taken as axiomatic that this has been a development pioneered by Western Civilization: it has spread over the whole world in the course of the last three centuries, so that the West can be regarded as providing the engine of world technological revolution since about 1700. It follows from these assumptions that the subject of this study is one of crucial significance both to the interpretation of recent history and to the understanding of the problems of the contemporary world. The concept of technological revolution thus gives a cohesion to the period since 1700 which is worth exploring and analysing.

Part Two

SOURCES OF POWER

3. The Ascendancy of Steam Power

If it is accepted that technology is concerned with the study of human techniques for making and doing things, an understanding of power is at the heart of technology because it is this which provides the ability to make and do anything. Sources of power thus figure prominently in the history of technology. At the beginning of technological development, the only source of power was human muscle, and the ability to make and use artefacts was constrained by the limits of human strength and dexterity. Gradually, other sources of power were exploited to supplement or replace the power of human limbs. Simple levers like the digging stick and the wedge were devised to increase the application of human muscle. Animals were harnessed to carry loads and to propel heavy instruments such as the plough, and it was realized that natural sources of power – wind, water and the tides – could also be used to perform certain laborious and repetitive functions like that of grinding grain. Such a realization, however, involved a grasp of basic mechanical operations and a range of constructional techniques in order to build a sound mill and equip it with appropriate gears and grinding stones. So the successful application of any new understanding of power sources could be delayed indefinitely by an inability to convert this power into a usable form. Some degree of technical expertise is essential for transforming knowledge into a manageable source of power. The technology of power, in short, is concerned with techniques of energy con-

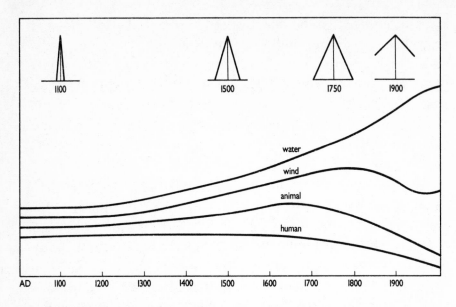

Figure 4. Energy gradients and the use of natural sources of power in Western Civilization

version. We live in a world surrounded by a vast flux of natural energy, derived largely from the sun as direct light or as movements of air and water generated by the sun, or as solar energy locked in carboniferous rocks in the form of fossil fuels. The primary task of human technological ingenuity has been to find ways of converting this energy into forms in which it can be used and controlled by mankind.

By 1700, substantial progress had been made in Western Civilization towards harnessing natural sources of energy, but human and animal muscle remained overwhelmingly predominant in most agricultural and industrial applications. For all practical purposes, this was what Fred Cottrell has called a 'low-energy society', depending mainly on animate 'energy converters' with a very high 'energy gradient' – that is to say, the power which they generated had to be consumed within the immediate vicinity of its source.[1]

44

Such societies tend to be characterized by high social stability and resistance to change, and this was true of Europe up to the seventeenth century, even though important signs of change had begun to appear in the Middle Ages and had grown distinctly stronger with the intellectual and religious revolutions of the Renaissance and Reformation, of which the birth of modern science was one of the most significant consequences. Other agents of change included the rapidly developing mercantile economy constructed upon the exploitation of the sailing ship, the first of the 'high-energy converters', which was capable of carrying goods and people over large distances for little additional input of effort after the initial fitting-out of a seaworthy vessel. The sailing ship had been invented in the ancient world – representations of such vessels appear in early Egyptian inscriptions. It was the first successful harnessing of an inanimate source of energy. Much later – probably in the twelfth century so far as Western Civilization was concerned – wind power was also applied to milling and other industrial uses in the form of the windmill, and windmills were certainly operating in many parts of Europe in 1700, especially in low-lying territory like the Netherlands, where they performed an essential function in land drainage.

By this time, however, wind power had been overtaken by water power in any agricultural or industrial function for which human or animal power had become inadequate. Water-wheels had been known since the Roman Empire at least, but they had undergone little development before the European Middle Ages. Since then they had proliferated on most of the rivers and fast-flowing streams of Europe. They remained almost invariably small and simple, using wooden undershot water-wheels (although horizontal wheels were preferred in some parts of Europe), with the minimum of gearing for power transmission. But great ingenuity had been shown in applying them to industrial processes such as fulling cloth, crushing mineral ores, sawing wood and operating machines. By 1700, moreover, overshot arrangements – whereby the water flowed into buckets on the top of the wheel rather than impelling vanes on the bottom of it – were becoming common; these gave greater efficiency than the undershot pattern. Wood remained the normal construction material.

These essays in wind and water power were important, not least because they demonstrated what could be achieved in terms of increased productivity and wealth creation, and prompted inventors and entrepreneurs to seek further improvements. But in quantitative terms their influence was small in scale compared with the overwhelming social reliance on human and animal power for most everyday purposes in agriculture, transport and domestic duties, so that Europe in 1700 was still a low-energy society. This situation was changed by industrialization, which required huge increases in energy in the space of a few decades and which promoted a series of technological innovations in order to make this power available. Three criteria came to be regarded as essential in any source of power developed to meet these novel demands. First, it had to be available in volume, so small-scale or marginal energy-converters were not encouraged at this stage. Second, it had to be reliable, as the drainage of mines and the smooth running of factories came to depend utterly upon it. And thirdly, the power source had to be reasonably accessible, both physically and intellectually: it needed to be easily obtainable, and capable of being moved, reassembled and maintained without insuperable difficulties. Several new sources of power were developed over the next three centuries to meet these criteria.

The transformation to a high-energy society is thus associated with the invention of new prime movers like the steam engine and the internal combustion engine, but before these novel means of energy conversion became available great steps were taken to increase the efficiency of windmills and water-mills. The nine-teenth-century British windmill, for example, with its easily con-trolled patent spring sails and its self-regulating fantail mechanism and centrifugal governor, was as much a product of advancing industrialization as the contemporary textile factory was, while water-wheels underwent significant improvements in performance as a result of metal construction and careful attention to water flow and regulation. Even without the steam engine, industrialization could have achieved much through wind and water power, and both continued to play an important part in the process throughout the nineteenth century. Yet it was the steam engine which provided the most spectacular demonstration of a new style of energy converter, capable of transforming Western Civilization into the world's first high-energy society.

The steam engine emerged in Western Civilization at the end of the seventeenth century. It had been anticipated in antiquity by the 'aeolipile' and other devices of Hero of Alexandria which demonstrated the principle of the steam-reaction turbine, but these were regarded at the time as little more than toys and they were not developed. With the stirrings of modern science in the seventeenth century, however, thought turned towards means of promoting the dominion of man over nature by increasing his power to make and do things. Two unexpected discoveries directed attention to the possibility of creating a new prime mover from the 'elastic' properties of steam: first, the revelation that the atmosphere pos-sessed weight, which varied according to the height above sea level; and second, the discovery that, despite the insistence of ancient authorities that nature abhorred a vacuum, it was possible to create a partial vacuum, either by the use of an effective air pump or by condensing steam in an enclosed vessel. It is not possible to be certain about the extent to which the early steam

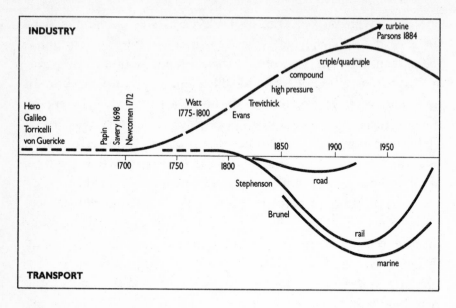

Figure 5. The development of steam power

engineers were familiar with these new scientific principles, as men like Savery and Newcomen were not acknowledged members of the scientific community and biographical details about them are sparse. It is clear, however, that they were working in an environment in which such knowledge was becoming easily available to people of intelligence, literacy and enterprise, and as the promoters of the first steam engines possessed all these qualities there is a fair inference that they had been stimulated by scientific speculation.

The first effective steam engine was that built by Thomas Newcomen to pump water out of a coalmine at Dudley, in the English Midlands, in 1712. Newcomen was a merchant in iron goods from Dartmouth in Devon. We know that he was a dissenter in religion, that he was literate and that he travelled around the country – all qualities which encourage the assumption that he would have encountered discussion of steam power and the possibility of creating a vacuum such as had been made general knowledge in the previous two decades by the publications of Denis Papin and other investigators. The Dudley engine was a substantial machine, built into its own engine house. The operating mechanism consisted of an upright brass cylinder, closed at the bottom and open at the top, in which a piston was free to move. Steam was admitted from a boiler below the cylinder and when this was condensed under the piston by a jet of cold water a partial vacuum was created so that atmospheric pressure on the top of the piston would drive it to the bottom of the cylinder. By connecting the piston to a large swinging lever or beam, this downward movement of the piston could be converted into an upward movement of pumping rods at the other end of the beam, so that water could be raised from any depth which the rods could reach. The weight of the pump rods was then sufficient to restore the piston to the top of the cylinder ready for the next stroke of the engine.

Several refinements were necessary to make this simple but cumbersome machine work smoothly, and Newcomen provided these in the shape of valves to control the inlet of steam and other functions, operated automatically by a rod from the swinging beam, and by devices for sealing the piston in the cylinder to make it airtight. Once constructed and coaxed into action, the Newcomen engine proved itself to be a robust and reliable machine, and it immediately supplied one of the urgent needs of an expanding industrial society, a good pump to make deep mines workable. As a native Devonian, Newcomen was almost certainly inspired by the need for such a pump in the tin- and copper-mining districts of south-west England. There was a problem, however, because the engine required abundant coal to provide the fuel for its boiler, and coal was a scarce commodity in that part of the country. The Newcomen engine was therefore rarely used in metal mining, but in the coalmining districts it was another story. Here coal was relatively abundant, and – an additional advantage – the boiler could consume the small coal which would otherwise have been regarded as waste by the proprietors. So the Newcomen engine thrived in the coalfields of the Midlands and the Scottish Lowlands, and especially in the flourishing coalfield of north-east England around the Tyne and Wear estuaries, which had already established a dominance in the London market because of the facility these rivers provided for sea transport to the metropolis. By the middle of the century there were about a hundred Newcomen-type engines at work in this area alone, and there can be no doubt that the steam engine was already playing a crucial role in the national economy of Britain by fulfilling the ever-mounting demand for coal.

Meanwhile, other uses appeared for the Newcomen engine. It was an efficient water pump, and one was installed in a London waterworks in the 1720s, although the high fuel costs here caused it to be discontinued. Even though the expense of coal dissuaded the metal miners of Cornwall and Devon from using it, the Swedish entrepreneur Marten Triewald took the idea home to build the first steam engine outside Britain, at the Dannemora iron mine.[2] German miners for metals in the Harz Mountains and elsewhere were also anxious to adopt the new technology and parties of British engine builders undertook to erect them. Amongst such itinerant millwrights, the Hornblower family from Cornwall was one of the best known, and it was one of this family, Josiah Hornblower, who travelled to America in 1753 to build the first steam engine in the New World, at a copper mine in New Jersey. Hornblower stayed on to build other steam engines for waterworks at Philadelphia and New York.[3]

The new technology thus spread steadily throughout Western Civilization in the course of the eighteenth century, wherever men of enterprise were able to recognize its potential and to find the resources of capital and labour to erect the bulky hardware. This process of transmission was not significantly assisted or retarded by patent legislation, in so far as it existed at this date. It is true that Newcomen felt obliged to seek protection for his invention under the very general terms of the patent granted to Thomas Savery in 1698 'for Raising Water and occasioning Motion to all Sorts of Mill Work by the Impellent Force of Fire', subsequently extended until 1733. Savery's invention had been for a device which worked by condensing steam in an enclosed vessel, thereby drawing up water from a depth limited by the ability of atmospheric pressure to raise a column of water, and then forcing the water higher by applying steam to the vessel at considerable pressure. Although it was used in some limited practical applications, particularly in circumstances where only the first part of the cycle was required (that relying on the condensation of steam and atmospheric pressure), the Savery engine was never adopted successfully by the mining community for which it was intended.

Nevertheless, Savery's success in acquiring a patent made it diplomatic for Newcomen to come to terms with him, so that the protection was extended to cover the new engine, and when Savery died in 1715 a syndicate of London businessmen continued to promote it and to receive royalties on all sorts of steam engines constructed in Britain until the patent lapsed in 1733. British patent law was slowly adapting itself to the novel concept of protecting individual property in inventions, and the case of Savery's patent was an important precedent in this respect. It was to become in time one of the most significant controlling factors in the promotion of a social context conducive to invention, but at the beginning of the eighteenth century the protection provided was still rudimentary, and in other parts of Europe it was even more primitive.[4]

By the second decade of the eighteenth century, the steam engine had thus been launched upon its prodigious development. For most of the first century of its existence, it remained predominantly an atmospheric engine, in the sense that the power came from the weight of the atmosphere responding to the creation of a partial vacuum by the process of condensing steam. Indeed, for much of this time it remained in common parlance a 'fire engine', because to the uninitiated observer it was the 'impellent force of fire' which appeared to be providing the power. By the end of the century, however, cautious experiments were beginning to make use of the expansive property of steam, whereby the steam engine underwent a further dramatic evolution. This development began with James Watt who, although not the inventor of the steam engine, is the man rightly regarded as the outstanding contributor to the technology of steam power. Watt was a Scottish craftsman and scientific instrument maker who, while employed by the University of Glasgow on the maintenance of laboratory equipment, applied himself to the problem of making a model Newcomen engine work efficiently. The result of his inspired meditation on this question was the invention of the separate condenser whereby, instead of the cylinder being alternately heated and cooled on every stroke of the piston, it became possible to keep the cylinder permanently hot while condensing the steam in a separate vessel, the condenser, which was kept permanently cool.

Watt made a model of the separate condenser and took out his first patent to protect it in 1769, but he experienced formidable difficulties in manufacturing a full-scale engine incorporating the new device and was not completely successful until 1775, when he entered into partnership with the Birmingham industrialist Matthew Boulton. In addition to essential capital and managerial expertise, Boulton provided a workforce of skilled artisans who were able to produce the finely machined parts which were required in order to achieve the full benefits from Watt's invention. The partners benefited from John Wilkinson's new method of boring cast-iron cylinders, and they were able also to obtain an extension of Watt's crucial patent until 1800, so that for the last quarter of the eighteenth century they established a virtual monopoly over production of the improved steam engine. In this period they produced and sold some five hundred machines, so it is not fanciful to see in this achievement the creation of the modern engineering industry, manufacturing complicated machines to high standards of precision and on a large scale for distribution to extensive markets. It is, indeed, true to say that when Watt came to the steam engine it was a cumbersome piece of hardware but he developed it into a fine product of precision engineering.

During the twenty-five years of the Boulton and Watt partnership, Watt made further substantial improvements to the steam engine. Of most general significance, he effected the transformation of the 'atmospheric' or 'fire' engine into a genuine steam engine by applying the expansive power of steam directly to the piston. He did this in the first place because, with the closing of the cylinder in order to keep it at a consistently high working temperature, it became necessary to substitute something for the weight of the atmosphere on top of the piston, and this could be done by admitting steam and allowing its expansion to assist the downward working stroke of the piston. Once this had been done, it was a comparatively simple step to admit and condense the steam alternately above and below the piston, converting what had previously been a single-acting machine into a double-acting one. This made the engine more efficient and smoother in its motion, so that it then became possible to adapt the machine for rotary action: that is to say, the simple reciprocating motion of the piston, which could be applied directly or through the swinging beam to the motion of pump rods in any water-pumping installation, could now be converted through a crank to provide rotary action and thus to drive any sort of machine depending upon the operation of wheels or rollers. The normal arrangement was to place the crank on the main axle of the flywheel, which became a feature of all double-acting steam engines in order to maintain the smooth running of the machine. Here Watt encountered a difficulty, because another inventor had taken out a patent for a crank applied to a steam engine and, even though it was doubtful whether the legal validity of this patent could be sustained, Boulton and Watt were anxious to avoid litigation which might draw attention to their own patent rights. Watt therefore used great ingenuity in devising alternative modes of converting reciprocating into rotary action, particularly his 'sun and planet' gearing which was adopted for all the rotary machines produced by the firm. This consisted of a cogged wheel (the 'planet') attached to the end of the connecting rod from the beam, which was fixed so that it ran around the circumference of a similar wheel (the 'sun') fixed on the axle of the flywheel: the 'planet' then transmitted the stroke of the piston very effectively to the 'sun' and turned the flywheel.

Two further innovations completed Watt's refinement of the steam engine. One was the 'parallel motion', an arrangement of swinging rods for keeping the movement of the piston per-pendicular even though it was attached to the end of a beam moving through the section of an arc. Watt was particularly proud of this invention because of its geometrical elegance, and it became a standard feature of all beam engines. The other innovation was the application of the centrifugal governor to the steam engine. The basic principle of this device, using weights spinning off the main drive of the engine to control the admission of steam and thus to regulate the speed of the engine, was not new and we have already noted its application in windmills. But Watt adapted it cleverly to control the speed of his rotary-action steam engines, and it became a standard fitting of all such engines thereafter.

The principle of rotary action had already been achieved in the primitive steam turbine of Hero, much more directly than by converting the reciprocating action of a conventional steam engine. Eighteenth-century inventors, including James Watt, were well aware of this precedent, and strove to replicate it in a full-sized steam engine, but they had no success. There were several mechanical problems, of which the most serious as far as Watt was concerned was the need to generate steam at considerably higher pressure than that which he normally used. All eighteenth-century steam engines, apart from a few of the very early devices of Thomas Savery, operated with steam at little more than atmospheric pressure, and considering the uncertain quality of boiler technology at the time this cautious policy was fully justified. The direct production of rotary action from steam power thus had to wait until the steam turbine was developed by Charles Parsons and others at the end of the nineteenth century. By this time, steam engineers had become familiar with the use of steam at high pressures and the construction of steam boilers had undergone a corresponding improvement.

Protected by the original patent of 1769, Boulton and Watt worked vigorously and ultimately successfully to recoup their heavy initial outlay of capital and expertise in the improved steam engine. Demand for engines was intense, and as most of them had to be constructed on site by agents of the company, the firm devised a system of payments based on the amount of work performed by the machine which eventually became an issue of great ill-feeling between themselves and their clients. Nevertheless, the engines were widely adopted, in Britain and abroad, with particular concentrations in south-west England, where their superior efficiency made them acceptable in situations in which the heavy fuel consumption of Newcomen-type machines was prohibitively expensive, and in London, Birmingham and Manchester. The partnership was thus outstandingly productive and significant, demonstrating the value of their steam engine in all sorts of industrial applications and familiarizing a wide range of customers with its possibilities. Despite the criticism of some of their contemporaries and subsequent historians that the tight control of their patent rights restricted the development of steam technology in the 1790s, when more daring innovators were anxious to experiment with high-pressure steam and steam locomotion, they deserve full credit for establishing the steam engine as a versatile prime mover.

However, when the Watt patents lapsed in 1800 the way was open for a series of dramatic developments in steam technology, most of which had been either actively discouraged or at least not positively encouraged by Boulton and Watt. The first of these was in the use of steam at much higher pressures than those normally used by Watt. The great advantage of this, once operators had acquired confidence in the improved boilers which became available, was that the expansive power of steam could be used more effectively, and consequently more efficiently. Oliver Evans experimented successfully with high-pressure steam in the United States, while the British pioneer of high-pressure working was the Cornish engineer Richard Trevithick. The large beam pumping engines to which Trevithick applied it, using the expansive power of a small amount of high-pressure steam against the partial vacuum produced by condensing steam from the previous stroke on the other side of the piston, became known as 'Cornish engines'. They were renowned in the first half of the nineteenth century for their superior reliability and efficiency, and they were installed for heavy-duty pumping and winding operations in metal-working mines throughout the world.

Another advantage of using steam at high pressure was that it became possible to make powerful engines which were much smaller and more compact than the conventional beam engine, and this opened up the possibility of building viable locomotive engines. William Murdoch, a gifted inventor and the agent of Boulton and Watt employed by them to erect engines in Cornwall, had already demonstrated that even a low-pressure engine could be made to propel itself, but it was a cumbersome device and he was discouraged by his employers from taking it further. It was left to Richard Trevithick to demonstrate that the bulk of a locomotive became manageable by using a high-pressure engine, and to produce the first practical steam locomotive for the Penydarren tram-road in South Wales in 1804. Although mechanically successful, this failed because the tram-road was not sufficiently strong to withstand the operation of the locomotives, but improvements in this respect soon followed and by the second decade of the nineteenth century the locomotive steam engine was being adopted for short-haul duties in collieries. By the third decade, longer lines were being constructed for steam haulage, and in 1830 the opening of the Liverpool and Manchester Railway signalled the beginning of the true 'Railway Age'. In this, the high-pressure steam engine, exhausting directly to the atmosphere without any condenser, played a crucial part and underwent continuous improvement in design to achieve higher speeds and better performance until it was replaced by other even more efficient engines in the middle of the twentieth century.

In addition to the introduction of high-pressure steam and the locomotive steam engine, every aspect of steam power underwent almost constant improvement in the nineteenth century. This process was stimulated both by the general recognition of the value and versatility of the steam engine, and by the severe competition of a more or less open market in which every saving in fuel consumption and advantage in performance could be exploited. There was no lack of inventive refinements to press engine design to ever greater efficiency, so that engine builders and engine users were constantly on the lookout for improvements from which they could profit. The shape of the steam engine broke away from the characteristic beam-engine style of the eighteenth century and assumed a bewildering proliferation of forms, with the cylinders placed vertically, horizontally or diagonally; with all moving parts made of iron or steel; and with valves developing from simple slide valves to, for example, Corliss valves, which were designed to give precise control and immediate cut-off.

Most spectacularly, the introduction of 'compounding' in the middle of the century, passing steam at high pressure two or more times through cylinders working at diminishing pressures, led to many new shapes: 'tandem-compound', with the high-pressure cylinder placed behind the low-pressure one; 'cross-compound', with the cylinders placed parallel to each other; and 'inverted-vertical', with the cylinders arranged upright and side by side and with their pistons driving downwards to a common crankshaft, being amongst the more common. Towards the end of the century, the need for high-speed engines to drive electricity dynamos and the competition of the internal combustion engine promoted further innovations such as the enclosed forced-lubrication engines capable of running indefinitely at high speeds, and the 'uniflow' engine in which heat losses caused by reversing the flow of steam on every stroke of the piston were overcome by feeding it from both ends and out through ports in the middle of the cylinder. By this time, however, the reciprocating steam engine was beginning to lose ground to the steam turbine, and the more visionary inventors were already turning their attention to other things.

Thus, as far as the production of power for all sorts of industrial and transport applications was concerned, the nineteenth century was dominated by the steam engine. This is not to say that other prime movers were rendered completely obsolete: water power remained very important in some areas and in certain processes, and the millwrights responded to the pressure for increased efficiency with skilfully designed water-wheels and, ultimately, with the water turbine – in several different forms – which continues to do valuable work in the generation of hydroelectric power. Tide mills and windmills also remained in use throughout this period, although their role diminished in importance. But it was the steam engine, as improved by Watt and provided with high-pressure steam by Trevithick and Evans, which captured the imagination of engineers and met the market demand for an efficient and versatile prime mover. By the middle of the century, the reciprocating steam engine was firmly installed as the main provider of industrial power; as the source of pumping power for mines and waterworks; as the means of locomotion on the railways of the world; and as the propulsive force of the steadily swelling armada of steamships which was already beginning to displace wind power from its last and most cherished bastion – the sailing ship.

This was an achievement of immense but incalculable economic significance, and it was also one of great imaginative impact. The harnessing of steam had been humankind's most daring acquisition of the Promethean fire, and it opened up vistas of boundless wealth and endless progress. Artists such as J. M. W. Turner were fascinated by the theme, as in his 'Rain, Steam and Speed', and Tennyson, Kipling, and others used steam-engine images in their poetry. The impact of steam engines on the landscape, through railway locomotives and factory chimneys, was profound, and although some commentators observed these developments with horror as depicting the encroachment of the 'dark satanic mills', the general mood of nineteenth-century opinion in Western

Civilization was one of optimism about the wealth and other benefits which the steam engine would bring. These expectations were not to be entirely disappointed.

4. *Internal Combustion and Electricity*

Even though the reciprocating steam engine dominated virtually every aspect of industry and transport in the nineteenth century, it was clear by the end of the century that its hegemony of technological power was being strongly challenged by other forms of energy conversion. First, the steam turbine had already demonstrated that it could perform some functions better than conventional reciprocating designs. Then the internal combustion engine began to exploit vast new potential markets for which the steam engine could provide no convenient alternative and, in the process, was to develop forms which were able to compete directly with steam in well-established usages. Third, the widespread availability of current electricity was undermining the monopoly of the steam engine in industry and transport. (Electricity is not itself a prime mover, as it requires mechanical generation to make it readily available, and this has normally been supplied by water or steam turbines. But its effect in application was to present a very serious challenge to conventional steam technology.) Finally, in the twentieth century, a new source of power derived from nuclear fuels was discovered, but for most practical applications this still had to be converted into electricity by steam turbines. The development of all these alternative sources of power made the reciprocating steam engine practically obsolete, and promoted a continuing revolution in power technology.

The basic principles of the steam turbine, involving the direct propagation of rotary motion through the impact of a jet of steam on the moving body (the rotor) or through the reaction from a jet of steam expelled from the rotor, had been known from antiquity. The first type, resembling the operation of an undershot water-wheel, is known as an 'impulse' turbine. The other type, working on the same principle as a rotating garden sprinkler, is a 'reaction' turbine. However, the problems of converting these principles into steam-powered units of significant size were considerable. For one thing, they depended upon a constant supply of high-pressure steam, and as we have seen the eighteenth-century pioneers of steam power were unwilling or unable to provide this because of inadequate boilers. When high-pressure steam did become available, it was found difficult to harness the energy from a jet of steam effectively until the water turbine had developed sufficiently to provide an understanding of this mechanism. By this time, moreover, the reciprocating steam engine was in its impressive prime, creating a high standard of performance which any new technology had to challenge. The development of the steam turbine was consequently delayed until all its operating problems could be solved, and this moment did not arrive until the last two decades of the nineteenth century.

The man who developed the first viable steam turbine was Sir Charles Parsons. His patent of 1884 was for a reaction turbine in which the steam passed through a series of rotors on the same axle separated by a series of fixed blades: the size of both moving and fixed parts increased along the length of the turbine to correspond with the reduction in the pressure of the steam as it expanded, and the flow of steam through the turbine set the rotors spinning at high speed. As with the water turbine, the steam turbine was a fully enclosed vessel, except for the points of entry and exit of the steam, and the flow of steam through the vessel was assisted by a condenser which drew off the exhaust steam.

Parsons coupled his first turbine directly to an electricity generator to demonstrate that the high speed of rotation was ideal for this purpose. The development of this machine for commercial application in the burgeoning electricity supply industry took place remarkably quickly, so that steam turbines were coming into widespread use in electricity supply stations by the end of the century.

By this time steam turbines had undergone a substantial increase in size and efficiency as Parsons and other inventors refined their mechanism. When he encountered patent difficulties in the use of his original 'axial flow' arrangement, Parsons experimented with a 'radial flow' version, with the steam flowing from the centre to the periphery through concentric rings of rotors and fixed blades. He showed thereby that the same operational principles applied, although with more mechanical complications. The Swedish engineer de Laval produced an effective design for an impulse steam turbine in 1889, high-pressure steam being directed through a carefully designed nozzle, the increasing bore of which converted the pressure energy of the steam into velocity and produced a very high speed of rotation. The rotor consisted of a wheel with curved blades on its rim into which the nozzle directed the steam, and was enclosed within a vessel from which the used steam was exhausted through a condenser. De-Laval-type turbines had to run at very high speeds because all the energy of the steam was utilized in a single stage, and they had to be greatly geared down in order to work effectively. Subsequent impulse turbines were devised which split the expansion of the steam into several stages, giving efficiency at lower speeds.

The steam turbine thus underwent rapid development, and this process continued for several decades. The small Parsons unit of 1884, generating 10 h.p., has grown into the giant turbo-alternators of modern power stations, developing some 600,000 h.p.

at a fraction of the fuel consumption of earlier steam engines. The efficient generation of electricity has not been the only achievement of steam-turbine technology. Parsons was quick to realize that, with adequate gearing down, his turbine could be a most effective source of propulsion for ships. When the British Admiralty was reluctant to admit this potentiality, Parsons converted them – and marine engineers in other parts of the world – by an unsolicited but sensational demonstration at the Jubilee Naval Review in 1897, when his experimental turbine-propelled launch the *Turbinia* cavorted amongst the comparatively sluggish naval ships at the incredible speed of 34 knots. Within a decade, steam turbines were being extensively adopted for propulsion for both naval and merchant ships, with the Cunard *Mauretania*, launched in 1906, being equipped with turbines which developed 70,000 h.p. and a speed of 27 knots. Steam turbines were eventually displaced by diesel engines for most purposes of marine propulsion, but they remain in use in some larger vessels to the present day.

Like the steam turbine, the internal combustion engine had antecedents long before it developed into an effective and successful challenge to the reciprocating steam engine at the end of the nineteenth century and into the twentieth century. Scientific commentators had observed that the action of a gun was essentially that of an engine, burning fuel in the form of gunpowder in a cylinder – the barrel of the gun – in order to produce movement in the projectile, which behaved like a piston in a steam engine, although it was a single-action motion that could not be readily repeated without reloading the fuel and ammunition. The non-repeatability of the action was a crucial disadvantage and prevented the development of a genuine gunpowder engine, but the conceptual possibility of such an 'internal combustion' engine, with the fuel being burnt inside the working cylinder rather than outside it as in a steam engine, became firmly established, and

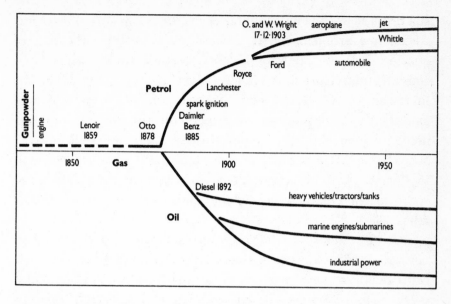

Figure 6. The development of the internal combustion engine

inventors were alert to the potential of such an engine once an appropriate fuel became available.

The fuel which enabled the internal combustion engine to become a practical proposition was coal gas. This had first been distilled from coal by William Murdoch, the agent of Boulton and Watt, in the 1790s, and it had been used to illuminate the Soho establishment of the Birmingham firm in 1798. Thereafter, its production and use spread rapidly, so that by the middle of the nineteenth century virtually every town of any size in Britain, and many in continental Europe and America, were equipped with a supply of coal gas. In 1859, the French engineer Lenoir succeeded in making an engine which ignited coal gas with an electric spark in a horizontal cylinder, thus driving a piston and using a flywheel to return the piston to the firing end of the cylinder. The machine was noisy, uneven and cumbersome, but it worked, and with sufficient development it could be transformed into a viable engine. The major problems in its early stages were the difficulties of igniting the fuel at exactly the right moment, and of achieving a smooth action from the violent nature of the working stroke. The first problem was tackled by substituting various arrangements of external flames or hot tubes to provide the ignition, but improvements in chemical batteries and electrical equipment encouraged a return to the ignition spark, and eventually the neat device of the 'firing plug', capable of delivering precisely the right sort of spark at the correct point in the engine cycle, was universally adopted.

Smooth action was achieved by the adoption of the 'Otto cycle' devised by the German engineer N. A. Otto in 1876; the cycle consisted of the four phases of injection (of fuel into the cylinder), compression, ignition and exhaust. The new science of thermodynamics had demonstrated that the fuel would burn most effectively if it was compressed, so Otto allowed the piston to compress the gas on its return stroke after it had been admitted to the cylinder on the first stroke, when the piston was withdrawing from the cylinder. The third stroke, the only working phase of the four, occurred when the compressed gas was ignited and burnt, driving the piston outwards once more. The return of the piston on the fourth stroke exhausted the burnt fuel from the cylinder and prepared the way for the repetition of the cycle. This cycle has been adopted by most internal combustion engines; the simpler form of the 'two-stroke' engine was developed for motor bikes and other lighter duties. The Otto cycle is intended to occur within every working cylinder, but an engine works best on this arrangement when there are four or more cylinders coupled together, with each cylinder designed to be performing one phase of the cycle at any given moment. In an engine of this type, such as that of the modern automobile, there is always one cylinder providing a power stroke and thus the smooth running of the engine is assured.

Gas engines were able to compete successfully with steam power for many small and medium duties in the closing decades of the nineteenth century because of the widespread availability of coal gas and the practical advantage of not needing the elaborate boiler plant for raising steam which was indispensable to the steam engine. There was, however, no realistic possibility of competing with steam for locomotive purposes because of the fixed nature of the gas supply. This situation was revolutionized by the introduction of an alternative fuel for the internal combustion engine – oil and its derivatives. Natural bitumen and other oils from vegetables and animals had long been exploited for their illuminating properties in lamps, but the development of the internal combustion engine opened up a vast new market for them, and the systematic exploitation of oil reserves, with a giant industry devoted to refining the products of the oil wells, became essential features of twentieth-century industrialization. Oil fuels not only increased the versatility and convenience of the internal combustion engine, they also made it mobile, because a supply of petrol (gasolene), paraffin (kerosene) or diesel oil could be easily carried in a tank to supply the engine. The result was an astonishing expansion in the production of this prime mover which has deeply affected every aspect of modern technological civilization.

Gottlieb Daimler used vaporized petroleum, the most volatile product of the oil-refining industry, which had previously been regarded as rather dangerous waste, to drive an internal combustion engine in 1885. Shortly afterwards he fixed it to a bicycle to make the first motor bike. In the same year, another German engineer, Karl Benz, used a single-cylinder petrol engine to drive a three-wheeled vehicle, so creating the 'horseless carriage', the first genuine motor car. The idea was taken up with enthusiasm in other countries of Western Europe and North America. Henry Ford, the most successful of all the new generation of entrepreneurial inventors who flocked to the automobile industry, built his first motor car in 1896, and in the same year British motorists achieved a notable victory with the repeal of the 'Red Flag' Act which had penalized all forms of mechanical propulsion on roads. By 1903 Ford had set up the Ford Motor Company of Michigan, which flourished in the following decades. He was quick to adopt promising inventions, such as the self-starter in 1911, and new metals like the vanadium steel which he used in his Model T. For this vehicle, the 'Tin Lizzie', the most famous of all mass-produced cars, Ford installed the new moving assembly line in 1913, and constantly improved it thereafter. By 1923, production had risen to more than two million a year and the Model T commanded a world market. Ford's methods, which were widely adopted by other automobile manufacturers, transformed modern industrial organization as thoroughly as his products helped to change our way of life.

Much of the basic technology of the internal combustion engine was borrowed directly from that of the steam engine. In the manufacture of accurately bored cylinders and precisely fitting pistons, valves and other components, the inventors who developed the new technology were able to benefit from the long experience acquired in making steam engines, and many manufacturers found it comparatively easy to make the transition from steam to internal combustion in response to the changing market demands. This process of using essential basic experience from an existing technology can be seen as an excellent exemplification of the ratchet concept, whereby a new technology learns from one which is already flourishing and then displaces it.

In some respects, however, internal combustion was significantly different from steam technology, and this was particularly the case in the level of scientific understanding required of both inventors and manufacturers. The steam engine had been a product of the empirical tradition of practical millwrights, who were constantly seeking to improve performance by attention to the details of working experience. It is virtually certain, as we have seen, that the steam-engine pioneers were familiar with the crucial scientific principles of the vacuum and atmospheric pressure, and a few leading engineers like James Watt were Fellows of the Royal Society and had a keen awareness of contemporary science. But for the most part, steam technology derived little from scientific investigation, at least until the end of the nineteenth century, and it could be said that it did more for science than science did for the steam engine. This is because it was speculation about the behaviour of the steam engine, and in particular the way in which it transformed heat into work, that promoted the science of thermodynamics. The new science, which had been fully articulated by the middle of the nineteenth century, then became important to the development of the internal combustion engine. Eventually, indeed, it became essential, because the high-compression engine developed by Rudolf Diesel in 1892 was inspired by the thermodynamic principle that it should be possible to induce self-ignition in the fuel by intense compression. Diesel had to overcome formidable operating difficulties before his engine could become a commercial success, but by the second decade of the twentieth century it was being adopted for a wide variety of heavy-duty engines, in ships, tractors and omnibuses, and it continues to enjoy enormous success today.[1]

It was natural to suppose that, just as the reciprocating steam engine had been superseded in many respects by the steam turbine, so the reciprocating internal combustion engine would give way to a design capable of producing direct rotary action. In fact, however, the parallel was misleading because burning fuel could not produce exactly the same effect as a stream of water or steam, and the creation of a combustion chamber in which to burn the fuel required stronger materials to withstand the consistently high temperature than any available to the early internal combustion engineers. However, Frank Whittle had already gone far towards establishing the principles of a gas turbine engine while he was an apprentice with the Royal Air Force in the 1920s, and he took out a patent for this in 1930. Discouraged by official lack of interest and by financial problems, he abandoned work on the project until the approach of the Second World War stimulated governmental attention and it became possible to assemble a research team in 1936. The first experimental engine ran in 1937, and development proceeded then until the first 'jet' planes entered service in the closing stages of the war. By that time other teams were at work on the idea, and it was the Germans who first got jets into active service.

The gas turbine became a jet engine because in its basic form it produced power by ejecting a stream of spent fuel from the combustion chamber in which it was burnt. There was some rotary action, but this was only used to obtain the compression of the fuel before ignition. Subsequent designs, such as the turbo-prop, aimed at making fuller use of this rotary function in order to turn a propeller or in some cases, such as experimental cars and railway locomotives, wheels. Other experimental designs such as the 'ram jet' have aimed at replacing the rotary compressors by relying upon the natural compression of high-speed intake, but this use is likely to be limited to high-speed, high-flying aeroplanes. Noise and heavy fuel consumption have so far prevented most land-based applications of gas turbines. But their low weight-to-power ratio makes them ideal for aircraft, and they have been almost universally adopted for large military and civil aeroplanes.

Apart from the gas turbine, there have been some less successful attempts to convert the internal combustion engine to direct rotary action. One of these, the ill-fated Wankel engine, was introduced as an automobile power plant in some German cars in the 1960s. It had a rotary piston, a triangular section of a disc which was driven round by the expansion of burning fuel pressing against each of its three faces in turn. The design was compact and it worked reasonably well, but it had protracted development problems and proved to be unpopular with the public, so the manufacturers abandoned production of the car, even though the engine has been developed for other uses. The fate of this promising innovation is a salutary reminder both of the tremendous resilience and adaptability of the standard reciprocating internal combustion engine, and of the danger of predicting technological obsolescence. Those commentators who were convinced that the reciprocating engine had reached the limits of its development – as many were prepared to say of the steam locomotive in the 1950s – were proved to be wrong. The standard petrol and diesel engines, together with the gas turbine, have come to exercise an astonishing hegemony over transport on land, sea and air in the twentieth century.

Like the steam engine before it, the internal combustion engine is a heat engine, and it is worth noting in passing that there have been other forms of heat engine which have so far not enjoyed great success, but are not without potential for future applications. The most interesting of these is the hot-air engine. Invented early in the nineteenth century, it depends for its working stroke on the expansion of heated air against a piston in a cylinder. While one end of the cylinder is kept permanently hot by an external source of energy, the heated air is repeatedly transferred to the cold end after each stroke and then returned to be reheated for the next stroke. Robert Stirling, the Scottish clergyman who produced a workable engine of this type in 1816, added a heat exchanger which alternately cooled and heated the air as it passed through. The hot-air engine underwent considerable refinement and proved itself to be a valuable source of power for small-scale uses such as driving fans, but in larger applications it was not able to compete with the steam engine. Then, in the twentieth century, the introduction of electric power displaced it even from those small-scale uses in which it had flourished, and it declined almost to vanishing-point. Some enthusiasts remain, however, for the hot-air engine, and it is still possible that commercial uses will be found for it in space and medical technology.[2]

Another heat engine, one which has demonstrated spectacular qualities and potentialities in the twentieth century, is the rocket. From being little more than a toy, this evolved rapidly in the Second World War as a means of delivering a powerful warhead far beyond the range of the biggest guns. The technology of the V2 weapon – together with Werner von Braun and other engineers who had worked on it – was then appropriated by the USA, and was used to explore the potential of a rocket engine as a means of space travel. Carrying its own supply of liquid fuels, which were combined and burnt in a combustion chamber, the rocket generated sufficient power to propel itself without 'breathing' atmospheric air. This meant that, unlike conventional forms of internal combustion engine, it had the capacity to function in the vacuum of space beyond the Earth's atmosphere. Space engineers in the USA and the USSR have developed this property for the exploration of space, so the rocket engine has been responsible for one of the most dramatic aspects of twentieth-century technology.

Rather more down to earth, electricity has become an enormously important source of power for industrial and domestic purposes in the twentieth century. It has already been pointed out that electricity does not constitute a prime mover because it has to be produced from some other source, chemical or mechanical.

The first step towards obtaining a usable electric current was taken right at the beginning of the nineteenth century, when the Italian physicist Volta devised the 'voltaic pile', a stack of metallic discs separated by brine-impregnated paper which caused a chemical reaction and produced an electric current. Scientists in Europe and America experimented with this novel form of energy, and in 1831 Michael Faraday demonstrated the practical possibilities of the relationship between electricity and magnetism both for the mechanical generation of electricity and for an electric motor. The principle was simple enough: Faraday showed that a current was induced in a coil of wire rotated between the poles of a magnet, and alternatively that when a current was passed through such a coil, the coil was deflected or caused to rotate by the magnetic field. It was several decades, however, before the implications of these discoveries were fully worked out, because many problems of mechanical design had to be resolved before work could begin on the manufacture of a commercially viable dynamo, and only then did it become realistic to develop electric motors. This stage was only reached in the 1880s, when Thomas A. Edison in the United States and several European entrepreneurs started to establish networks of electrical power supply. Edison, in particular, saw the value of a reliable power supply as a means of marketing the incandescent filament lamp which he had invented, and once the network was in place it was available for other uses, such as providing power for a transport system using electric traction. Edison used direct current in his installations, but many other electricity generating enterprises in America and Europe preferred alternating current at a much higher voltage because it could be transmitted over greater distances, and it was the alternating system which was eventually adopted universally.

Once electricity became generally available, its immense convenience and versatility meant that it was applied to many activities in industry, transport and the home. In industry, compact power units on individual machines replaced the large steam engines operating through belts and shafting which had been customary in mills and workshops in the nineteenth century. As a direct result of this, industry became more dispersed, because the 'energy gradient' of electricity was extended almost indefinitely and manufacturers no longer needed to give priority to proximity to the source of power when siting their factories. In the transport field, the initial impact of electricity was more ambiguous because the electric tramcar became, after a brief period of great success, too cumbersome and inflexible to compete with the internal combustion engine in providing urban transport. On the other hand, electric traction quickly acquired a monopoly of underground urban transport services in the capital cities of the Western world, and they came to exercise a profound influence on the development of suburban commuter traffic. Eventually electricity also came to provide an attractive alternative to steam and internal combustion on high-speed inter-city railways such as the prestigious French TGV (Train à Grande Vitesse). But it has as yet hardly impinged on road transport, even though much attention has been given to the improvement of batteries capable of powering road vehicles with a standard of performance commensurate with that of the automobile.

As far as domestic applications are concerned, electricity has wrought a revolution which is so complete that it is virtually taken for granted in most homes in the advanced industrial societies. Light, heat and cooking facilities are all available now at the touch of a switch, as are the means of laundering, vacuum-cleaning and a host of other previously laborious household tasks, as well as all the familiar modes of mass communication and entertainment. Electricity in the home has thus profoundly influenced attitudes towards domestic service and the traditional responsibilities of

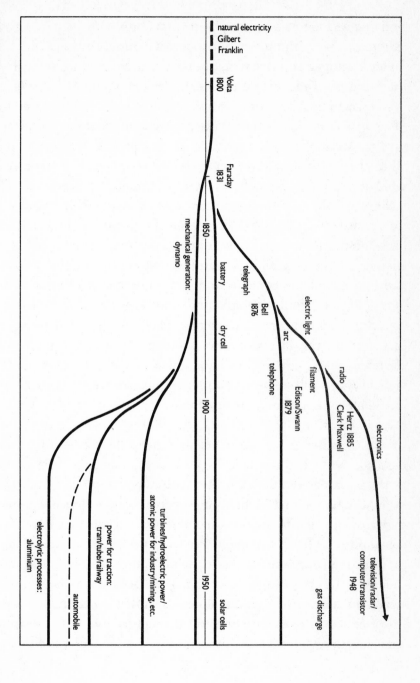

Figure 7. The development of electric power

men and women. It is not surprising that Lenin identified electricity as one of the components of modern Communism, nor that the underdeveloped countries look hopefully to acquiring electricity in their homes as a means of emancipation from domestic drudgery.

In reviewing the continuing revolution in sources of power since 1850, it is remarkable that there was such a cluster of significant innovations in the last two decades of the nineteenth century, since when there has been almost a century of comparative stability in which these innovations have been assimilated and matured. The technological breakthroughs achieved by Parsons and de Laval in steam turbines in the 1880s, by Daimler and Benz in the 1880s and Diesel in the 1890s in the development of the internal combustion engine, and by Edison and European engineers such as Siemens and Ferranti in the generation and application of electric power in the 1880s were all seminal in the sense that they sowed the seeds of technological progress for many decades into the future. The fact that, in the late twentieth century, we are still producing electricity largely by steam turbo-generators, and that our vehicles and ships are still powered mainly by reciprocating internal combustion engines, demonstrates the extraordinary importance of these innovations.

While the history of power technology in the twentieth century has thus been distinguished by its stability in so far as its basic processes have been concerned, that does not imply that there has been little happening. In fact, the process of assimilation has been pursued with vigour and resourcefulness, and some very important developments have occurred. For one thing, there has been a pronounced increase in scale. Much more power has been produced, by larger turbines grouped into fewer but more massive establishments. The economies of large-scale production have provided a strong incentive for this concentration and increase in scale. Secondly, design modifications have achieved continuing improvements in performance through changes in detailed arrangements, the use of better materials as these have become available and constant tuning to prevent waste and improve efficiency. This has been particularly notable in the spectacular success story of the reciprocating internal combustion engine – though it has been obliged to yield to the gas turbine for aviation. But it also applies to refinements in steam turbines and generators, to electric supply systems and to the design of electric engines. In virtually every aspect the networks of power in twentieth-century technology have been subject to continuing improvement in detail in order to maximize their operational economy.

Thirdly, even though the engines providing power to modern society have undergone little fundamental change in the twentieth century, there have been striking innovations in the fuels from which this power has been generated. Most important amongst these innovations has been nuclear fuel. Physicists discovered radioactivity in the 1890s and explored the fundamental structure of the atom in the following decades, but until the Second World War many of them remained sceptical about any practical application of this knowledge. Then the Manhattan Project and the race to build an atomic bomb made available the resources to build the first atomic pile and culminated in 1945 in the bombs dropped on Hiroshima and Nagasaki. The return of peace promoted attempts to develop more constructive applications of the energy locked up in the atomic nuclei of uranium and plutonium. This was achieved through the heat generated by controlled nuclear fission within an atomic pile, which could readily be converted into steam and thence, through conventional turbo-alternators, into electricity. Formidable design problems had to be overcome to ensure the reliability and safety of the new atomic power stations, but from the 1950s they became common adjuncts to the electricity generation grids which made power available to

77

consumers all over Europe and North America. Elsewhere they were less common, partly because of security considerations and the desire to restrict nuclear technology to the major powers, although this did not in fact prevent the construction of nuclear plant in several Third World countries. They have been most popular, however, in European countries, and particularly France, in which a shortage of fossil fuels has encouraged the use of nuclear fuels.

In the last decade, a series of accidents, especially that at Chernobyl in the USSR, have cast a shadow over the nuclear power programme, so that its future now seems uncertain. Critics of the programme have repeatedly observed that the industry has not yet worked out adequate means of disposing of its radioactive waste; now that the first generation of atomic power stations have fulfilled their working life and are being taken out of commission, the enormous costs of the decommissioning process are becoming apparent. It has also become clear that in such a complex establishment as a nuclear power plant, the possibly horrific consequences of mechanical failure or human error can never be entirely eliminated. So industrial society is pausing to reassess the implications of the nuclear power programme. Nevertheless, it can be argued that nuclear fuel provides a viable alternative source of power, that the world will eventually run out of fossil fuels and that nuclear power is 'clean', in the sense that it does not pollute the atmosphere or the environment with exhaust gases. The problem of disposing safely of radioactive waste at the end of the process remains, however, as a formidable obstacle to the extension of nuclear power.

Finally, it is worth noting that the nuclear power industry provides an important base from which the tantalizing prospect of power from nuclear fusion may eventually be attained. All conventional nuclear power stations consist essentially of atomic piles in which the heat from nuclear fission is generated and controlled. But if the atomic nuclei can be encouraged to fuse rather than to disintegrate, there will be an enormous release of energy – as in a hydrogen bomb – and with virtually no waste product. The raw material, moreover, should be derivable from sea water, without the intervention of dangerous radioactive materials. The theory of this process is feasible and elegant. The trouble is that in practice it requires a very high initial temperature, such as those involved in an atomic explosion, in order to start the reaction, and attempts to generate this sort of temperature in an enclosed and controllable environment have so far been frustrated. The most hopeful line of development seems to be that which depends upon the generation of a plasma of atomic particles held in a doughnut-shaped vessel by magnetic fields which prevent it from coming into contact with any solid surfaces. But even the most optimistic physicists are cautious about predicting any viable outcome from this experimental work until well into the twenty-first century.[3]

Meanwhile it seems likely that sources of power which have dominated technology throughout the present century – electricity produced mechanically from turbo-alternators and the internal combustion engine in its various forms – will continue to hold sway. They are, however, dependent on fossil fuels, which are a declining asset in terms of availability and cause an increasingly unacceptable level of atmospheric pollution. Alternative sources of power do not at present seem able to produce anything like sufficient energy to meet the intense hunger of modern industrial societies for this vital component of their prosperity. There is clearly scope for an increase in the production of power from natural sources such as the wind, the tides and water, and it should be possible to produce power from these sources much more efficiently than by the windmills and water-powered establishments which flourished in the eighteenth century. There is scope also for a significant increase in the use of solar energy, both in the direct provision of heat and in the form of power from solar cells. It is possible that prime movers which have hitherto been disregarded, like the hot-air engine, or those which have become obsolete, like the reciprocating steam engine, may be revived, because although both are heat engines their effect on atmospheric pollution is less harmful than that of the internal combustion engine. But none of these alternative sources of power seem likely to be available sufficiently soon or on a sufficient scale to avert the energy crisis which begins to loom ahead of industrial civilization as it reaches the end of the twentieth century. One way or the other, it is probable that the comparative stability of power technology enjoyed in the present century cannot be maintained much longer.

Part Three

APPLICATIONS OF POWER

5. The Emergence of the Factory

One of the most distinctive features of the modern world is the capacity of human societies to increase their productivity, the proportion of goods produced in relation to the size of the total population. In a literal as well as in a metaphorical sense, they have discovered the way of making two blades of grass grow where only one grew before, and they have thereby enormously increased the total wealth of society. This makes no reference to the justice or fairness with which the wealth is distributed between the members of society, but is a bald statement of the central fact of modern life: we have achieved an increasing rate of production of food and other material goods which has – so far at least – more than kept pace with the rate of population growth. And the main key to this achievement has been the availability of a stream of technological tools, from the new sources of power to machine tools and implements of mass production. In a very positive sense, therefore, it is an achievement of technological revolution.

The most important of all consumer goods is food, for the obvious but basic reason that without it human societies lose their vitality and starve. It is thus with good cause that the early steps to speed up the processes of food production have frequently been described as the 'Agricultural Revolution', although this is only meaningful within a wider context of industrialization, in which the production of food is merely a part, albeit a crucial part. The whole process of industrialization has come to be regarded familiarly as the 'Industrial Revolution', and even though there are many practical and conceptual problems about this term, particularly if it is applied too narrowly in relation either to time, period or place, it retains a certain rough serviceability as a description of the total ongoing process of industrial transformation with which we are concerned.

The Agricultural Revolution which began in Western Europe in the eighteenth century was, in the first instance, a political or administrative change rather than one in which technological innovation figured prominently. It was promoted by the inadequacy of traditional types of landholding and agricultural organization in response to the challenge of an increasing market. Open-field farming and a society based upon forms of serfdom had served the material needs of the European population well enough from the Teutonic settlements of the fifth and sixth centuries AD, but were unable to respond to the increasingly mercantile and expanding social needs of the seventeenth and eighteenth centuries. Change came either gradually and constructively, by the enclosure of farm lands and the movement of the population to the growing towns, as in Britain, or suddenly and explosively, as in the massive reallocation of land to the peasants in the French Revolution.

Elsewhere in Europe, serfdom was steadily abolished and replaced by more business-oriented forms of agricultural organization. Once this reorganization was in process, first in Britain and gradually thereafter in the rest of Europe, it became possible to introduce new technologies into agriculture. Mechanical innovations were at first modest, with improvements in ploughs and basic implements coming first, but with some significant inventions in machines for sowing seed and threshing. There was little scope for new sources of power until well into the nineteenth century, when determined efforts were made to introduce steam engines into British farms. These were particularly successful in the large farms of the English lowlands such as Norfolk, where techniques of steam ploughing were perfected. For this a substantial steam engine was required on the field to be ploughed, and as the weight of the machine made it impossible to wheel it across the field it was necessary to set up bases from which the plough could be hauled across the field by steel wire. The system was cumbersome, but it worked well where large fields were normal, and it was adopted in many parts of Western Europe by the end of the nineteenth century. There were then some experiments with forms of electric winding, but these and the well-established steam-ploughing techniques quickly succumbed to competition from tractors with internal combustion engines.

Other techniques were more important than mechanization in the early stages of the Agricultural Revolution, especially those concerned with the improvement of crops and animal stock. Methods of crop rotation were developed in the Netherlands in the seventeenth century which enabled most of the arable land to be kept under cultivation all the time, instead of putting it out to fallow once every three years. These methods were introduced into East Anglia and were readily adopted elsewhere, as were exotic new crops from the New World such as the potato. Farmers on the enclosed farms were now able to control their stockbreeding and were thus encouraged to experiment in raising new types of cattle, sheep, pigs and horses in order to meet particular demands of the expanding market for different types of food, wool and horse power. Men like the English Midlands farmer Robert Bakewell were able to make sensational progress in these respects, and their achievements were closely watched and emulated by the farming community throughout Britain. Their techniques were then taken up in New England and Western Europe.

The 'model farms' which became leading exponents of the new

farming practices in Britain in the middle of the nineteenth century were thus fully enclosed farms under a single owner or tenant-farmer who employed a variable team of wage-labourers to perform under his directions. They normally possessed substantial buildings, often arranged around a court equipped with a supply of water and a portable steam engine to provide power for the agricultural machines as required. The arable fields were well drained and regularly ploughed and harrowed, being planted every year on a carefully designed system of rotation. Pastures supported the farm's herds and flocks, which were chosen and bred with specific market objectives in view. Meadows provided hay for winter feed, and any woods on the estate were also farmed systematically. This type of farm was a direct product of the Agricultural Revolution, and in 1850 it was very popular and could be found in all parts of the country. Above all – and this was the reason for its success – it was profitable, and enabled British landowners and tenant-farmers to make a comfortable living from the land, at least until this type of farming came into competition with the much larger farms of the New World, from which grain and meat began to reach the British market in the last quarter of the nineteenth century. This brought to an end the prosperity of the 'high farming' years, and precipitated a long period of dislocation as British farming adjusted itself to the new situation. But this disturbance demonstrated the degree to which farming had been integrated into the process of industrialization in the Western world, and there was never any question of reverting to a more primitive form of agriculture.

Just as the production of food responded to industrialization, so did the mining and extractive industries. Some of these, like the metal-mining industries of Saxony and Slovakia, had a very long ancestry; the degree of sophistication which they had reached by the middle of the sixteenth century is vividly represented in the great work of the German mining engineer Agricola, *De re metallica*, with its many detailed woodcuts illustrating the tools and processes then in use. While markets remained fairly small and were restricted by transport problems, the production of these extractive industries was a rather peripheral activity of most human societies, and the miners were regarded as a remote body of men living largely under para-legal jurisdictions peculiar to their own special circumstances. But increasing demand and improved facilities for transport brought the mining communities and their activities steadily into a more integrated relationship with the rest of society; the general population came to depend more and more on what they produced, but also became more conscious of the waste and general despoliation of the environment which they generated.

This was especially the case with the coalmining industry. The use of coal as a fuel was one of the most distinctive features of early industrialization in the West. There are a few earlier references to the use of coal in Roman and Chinese civilizations, but for most practical purposes the serious mining of coal as a fuel began in late medieval Europe and expanded dramatically thereafter. The first mines were shallow excavations on sites where the coal measures outcropped on the surface, and it was rarely practicable to pursue the seams to any depth when plenty of coal was easily accessible near the surface. The typical arrangement in early coal-fields was thus one of lines of 'bell-pits' following the lines of the outcrops, the name deriving from the characteristic cross-section whereby the working of the coal spread out slightly at the bottom of a short shaft; the spoil from such mines was deposited in a ring round the mouth of the shaft, so that abandoned bell-pits leave a dimpled landscape of ringed banks and depressions.

Coal was first used abundantly as a domestic fuel, particularly in the growing towns of northern Europe, when timber began to become relatively scarce; by the sixteenth century it had become profitable to ship large quantities of it from the Tyneside coalfield to London. Coal was regarded as unsuitable for many industrial processes – such as the refining of metal – because its smoky by-products contaminated the materials being worked, but in other processes where it was possible to separate the production of heat from the materials being processed, coal began to have important industrial applications. In soap-making, for instance, and in the manufacture of glass, paper and ceramics, coal became an increasingly important fuel, and its availability acted as a powerful locational factor in the development of these industries. The remarkable concentration of the British fine ceramics industry in the legendary Five Towns of the Staffordshire Potteries is a case in point: it was influenced by the presence of local coal, suitable clay and the development of a convenient transport system.

By the beginning of the Industrial Revolution, therefore, coal was already established as an important fuel, and was steadily displacing wood fuels for industrial as well as domestic purposes. The advent of steam power increased the need for coal, so that one of the most significant indices of increasing industrialization became the rise in demand for and production of coal. This meant that the simple early mining techniques became inadequate to cope with the escalating demand, so a search for new sources of coal was begun and existing sources were more thoroughly exploited. Shafts were sunk deeper in search of workable coal seams, and the process was greatly assisted by the steam engine, which could pump unwanted water from the mines. The trouble was that coal shafts acted like wells, piercing the subterranean water table and collecting water; the technique of building them, indeed, had much in common with that for well-making. Without an efficient pump, therefore, coalmining was very severely constricted; conversely, with a steam-powered pump operating continuously, it became possible to sink shafts to previously inconceivable depths and to extract a much greater volume of coal from a single mine.

There were other technical constraints on coalmining as well as that imposed by unwanted water. Methods of cutting the coal, for example, remained surprisingly primitive until the twentieth century, with by far the greatest reliance being placed upon human muscle wielding picks and shovels. Transporting the coal and spoil underground, moreover, was a constant problem until mechanically operated trucks and conveyor belts were installed – again, mainly in the twentieth century. And raising the coal and spoil to the surface – and, of course, conveying the miners themselves safely to and from their work – depended upon the introduction of steel wire for the winding engines in the middle of the nineteenth century. But by far the most serious problem of coalmines was that caused by the potentially lethal explosive qualities of the coal gas and dust liberated by the mining process, which seriously restricted the illumination available to the miners at a time when all artificial light was derived from burning tallow or wax. Hence the enormous importance attached to the development of a safety lamp in which the burning wick could be separated from immediate contact with its potentially volatile environment. This technical advance was attributed variously to Sir Humphrey Davy and George Stephenson, and was undoubtedly a boon to the coalmining industry from the 1820s onwards.

Whether or not this and the other technical advances which we have considered should be regarded as boons to the miners themselves is more problematical. Coalmining communities grew quickly on the coalfields of Europe and North America in response to the tremendous increase in the demand for coal, which continued throughout the nineteenth century and reached its peak about 1914. They tended to consist mainly of dispossessed agricultural labourers and other immigrant groups, so they lacked the tradition of independence acquired over many centuries by the metal miners. They lived in poor housing, often isolated from other communities and without many of the amenities of a well-ordered town life. The conditions of their employment in a generally uncontrolled capitalist system meant that they were usually underpaid and exploited: even women and children were expected to work in the mines, until interventionist legislation began to prohibit such practices in the middle of the nineteenth century. The work was hard, dirty and dangerous, with frequent horrific accidents, and these were made more rather than less frequent by the technical improvements which increased the intensity of mining operations without commensurate attention to the safety and well-being of the workers. There can be no doubt that the spectacular increase in coal production was one of the most important single aspects of the whole process of industrialization, but from the point of view of those most intimately concerned with coal production – the workers in the industry – the benefits of this success were painfully slow in making themselves felt.

The traditional metal-mining industries made remarkably little direct technical contribution to coalmining, and tended to be overshadowed by the latter in the eighteenth and nineteenth centuries. But they remained crucially important to several regional economies, and the materials which they produced – tin, copper, lead, arsenic and other metals, including silver and gold – made a vital contribution to the processes of industrialization. Because they are mainly associated with older and harder rocks than the coal measures, these metals have usually been mined in highland areas well away from the main centres of population. In Britain, the tin and copper mines of Cornwall and Devon were typical in this respect, and had a distinctive jurisdiction – the medieval stannary courts – which reinforced their isolation from other communities. Here, rather than in the coalmines, the techniques of Agricola were copied and developed, including elaborate processes for crushing the metallic ores and refining them to extract the maximum amount of pure metal. Apart from the problem of disposing of unwanted water, which the industry solved with its own ubiquitous 'Cornish' steam engine, the difficulties of metal mining were rather different from those of coalmining. In particular, there was no special anxiety about explosive gases, but the hard rocks were much more difficult to work than the comparatively soft rocks of the coalfields. This led to the development of rock-drilling systems, culminating in pneumatic drilling driven by compressed air from a centrally placed compressor, and to the extensive use of explosives to dislodge the rock. From a social point of view, however, it has to be said that hard-rock miners fared little better than coalminers in the eighteenth and nineteenth century, and being generally in remote areas they tended to share in the comparative impoverishment of their regions.

One metal was of outstanding importance in the Industrial Revolution – iron. The ores of iron are widespread, being found in both hard and soft rocks, although inevitably it is the latter which have been most extensively worked, often by opencast techniques as in the ore fields of Lorraine. The biggest changes in the iron industry, however, apart from the enormous increase in the scale of operations, has come in the processes by which the ore has been worked to extract the iron, and in the ways in which the metal has been subsequently treated to produce different forms of iron and steel. The traditional 'direct' process for producing wrought or malleable iron from a small charcoal-fired hearth known as a 'bloomery' was transformed in Europe in the late Middle Ages by the introduction of the blast furnace, a substantial stone structure in which a high temperature was maintained for months on end in a charcoal-burning furnace by the use of water-powered bellows, and from which the iron was extracted in molten form to produce pig, or cast, iron. This is a hard but brittle form of the metal which requires further treatment in a refinery to convert it into wrought iron: it is consequently known as the 'indirect' process, in contrast with the malleable iron produced directly from a bloomery.[1]

The next significant series of changes came in the Industrial Revolution, and were associated with the conversion to coal fuels. Iron was first produced from a coke-fired blast furnace by the Quaker forge master Abraham Darby at Coalbrookdale in 1709: as the demand for iron expanded in the eighteenth century other entrepreneurs joined the search for good coking coal in order to follow his example. Charcoal continued to be used in some of the remoter furnaces in Europe and North America, but by the middle of the nineteenth century the transition to coke was virtually complete. The load-bearing qualities of coke made it possible to build larger blast furnaces, and the introduction of a preheated blast of air (the 'hot blast') further increased the productivity and efficiency of the industry. The transition to coal fuels also brought the blast furnaces literally from the backwoods to the coalfields, where they were stimulated by the presence of a larger market and labour force, and by proximity to other industries. Iron, especially cast iron, became universally available for a wide variety of structural purposes in fireproof mills and bridges, as well as for domestic purposes such as cooking implements. This easy availability of cheap iron was one of the most characteristic aspects of rapid industrialization in the nineteenth century.

Meanwhile, the other iron and steel processes were being transformed to meet the rising demands of the market. Cast iron could be reheated and refined by 'puddling', or stirring, to produce wrought or malleable iron, and this made available increasing quantities of high-quality material for bars, rails, plate and wire. Thanks to Henry Cort's new process, patented in the 1780s, it became possible to use coal fuels for puddling, in a reverbatory furnace, in which the hot gases pass over rather than through the charge; this meant that the refineries, like the blast furnaces, could move to the coalfields. And Huntsman's process for making crucible steel meant that this alloy of iron and carbon, mixed in carefully calculated proportions and sealed in ceramic pots, could also be made in a coal-fired furnace. The cementation process, for baking iron bars with charcoal to produce blister steel, was similarly adapted in the middle decades of the eighteenth century. These were not as yet mass processes for producing steel in abundance, but at least it began to become widely available for machine parts and other uses requiring special qualities of hardness and durability. Like the other iron processes, moreover, they promoted the convergence on the coalfields, with all the consequent stimuli to demand and productivity which followed. Thus by the middle of the nineteenth century a vigorous iron and steel industry based on coal fuels was transforming the economy of Europe: in the German Ruhr district, in France at Le Creusot, in Belgium around Liège, and most particularly in Britain – in the Midlands, south Yorkshire, south Wales and Lanarkshire – a tremendous quantity of iron goods were being produced of every conceivable type and quality. The Crystal Palace of 1851, the splendid edifice of cast iron and glass in which Britain was host to the world for the Great Exhibition, was a most appropriate symbol of the manufacturing and technical processes which had made modern industrialization possible.

The spirit of vigorous innovation which animated the iron and steel industries in this period was apparent in other productive industries also. We have already had cause to note developments in the glass and ceramic industries, and these represented a wide range of processes in which chemical reactions played a vital part, all of which responded to the demand of a growing market and to opportunities for new products. Some of these industries, such as the manufacture of pottery, were very ancient and had undergone little development in their basic technology – in this case, shaping of carefully chosen and well-mixed clays on a potter's wheel, and baking the artefacts so formed in an oven. Others, like glass-making, which achieved a high standard of technical excellence in Venice, from where it spread over the rest of Europe, were resumed in Western Civilization in late medieval times. Others again, like the manufacture of gunpowder, paper and brass (an alloy of copper and zinc), were almost certainly introduced into medieval Europe from China, although the lines of technological transmission remain obscure. High-quality porcelain was definitely an innovation from China, the result of patient emulation of Chinese specimens by Western craftsmen, who succeeded first at Dresden (Meissen) and subsequently in Britain, where William Cookworthy achieved his first success in the 1760s. By using kaolin (China clay) and firing it at very high temperatures he managed to secure the chemical change to translucent porcelain.

All these processes were, to some degree, chemical in nature, and they represented a steadily widening confidence in the ability of the industry to handle elaborate chemical changes. They were also stimulated by the processes of industrialization in the eighteenth century, so that they grew substantially in size, particularly in Britain, where the potential rewards for entrepreneurial innovation were greatest. It was in Britain, moreover, that the lead was taken at this time in the development of a bulk chemical industry, producing large quantities of acids and alkalis for use in other industries. These materials had hitherto been produced only on a small scale for laboratory purposes, usually by apothecaries or dyers. But in the mid eighteenth century the Scottish industrialist

John Roebuck patented his lead-chamber process for producing sulphuric acid on a large scale. One of the many industrial uses for this was in producing sodium carbonate, the alkali favoured by the soap-makers, and the French scientist Leblanc invented a process for mass-producing this 'soda'. But his efforts to develop the process in France were frustrated by the French Revolution, and it was in Britain that it first took firm root, especially on Merseyside, round Widnes, and on Tyneside. It was a noxious process, producing large quantities of unwholesome waste, and, until restraints were placed on such activities, the heavy chemical industry came to be associated with blighted landscapes.

Chemicals were required in large quantities in Britain in the eighteenth century by the rapidly developing textile industry. The woollen-cloth industry had been the bulwark of British industrial and mercantile prosperity from the Middle Ages, and had used chemicals for scouring, bleaching and dying the woollen fabrics, amongst other things. Most of these chemicals were natural, including human urine, alum and vegetable dyes, and as such had managed to keep pace with the growth of woollen-cloth manufacture. But the increased production of other textiles, such as silk, linen (from flax) and cotton, placed a heavy pressure on the traditional sources of chemicals, and thus helped to encourage the expansion of that industry. The popularity of the newest of these textiles, cotton, raised special problems of innovation and adjustment in eighteenth-century Britain. Cotton is a subtropical plant, and it had been brought to British markets as a result of the flourishing trade with the Middle East, including Egypt, where it was grown. The realization that it could be used to produce a light and durable fabric suitable for a very wide range of domestic and public applications promoted a rapid growth in the cotton-textile industry, which became so spectacular in the course of the eighteenth century that it came to epitomize the Industrial

Revolution. Generations of school textbooks saw this process beginning with a sequence of mechanical innovations which transformed the manufacture of cotton goods from a domestic industry into a large and fully integrated series of factory-based operations. For textbook purposes, these began in the middle of the eighteenth century with the invention of James Hargreaves's spinning jenny, which made it possible for the operator to spin many threads at the same time; Richard Arkwright's spinning frame and Samuel Crompton's spinning mule mechanized the process, using water power or steam.

A prime incentive for this acceleration of the spinning process was the fact that it traditionally took the work of a dozen or so spinners to keep one weaver busy, so that there was something of a bottleneck in the spinning department. However, the success of the new spinning machines was such that the bottleneck in production now shifted to the weaving and finishing departments, so that these too came under pressure to mechanize. First of all, an invention of the 1730s – John Kay's flying shuttle, which had hitherto languished largely unwanted because the weavers had no need to speed up – was taken up and exploited. By providing mechanical means for propelling the shuttle across the loom, Kay's invention enabled the weavers to work much faster; then Edmund Cartwright's power loom dispensed with the need for an individual weaver at every loom and enabled a battery of looms to be operated by water or steam power. Meanwhile, machines were introduced for preparing the raw cotton for spinning and for printing coloured patterns on to the finished fabrics as required, so that virtually the whole process of cotton manufacture was mechanized as the successive bottlenecks in production were eliminated.

The transformation of the British cotton industry between the 1750s and the 1830s is an outstanding illustration of a bottleneck sequence, because it shows how, in sufficiently favourable circumstances, entrepreneurial determination will shift from one process to another in the attempt to secure an overall increase in productivity. It is hardly necessary to add that this process is rarely seen so clearly or on such a large scale, because many factors of industrial inertia or socio-political intervention can easily disrupt it – not to mention a lack of initiative by the people involved. But, being a comparatively new industry, the manufacture of cotton at this time was remarkably free from constraints. The industry had also attracted some industrialists of outstanding quality who were ready to seize opportunities for maximizing production. The result was that it was converted within a lifetime from a small-scale industry practised in the cottages of the workers to a vast factory-based industry concentrated – once the dependence on water power was eliminated by the widespread introduction of steam power – in the towns of Lancashire and Lanarkshire. In these areas a totally new landscape came into existence: large mills with chimneys belching out black smoke, in towns made up largely of the terraced homes of the mill workers. It was not a pretty land-scape, but it was redolent with economic growth and dynamic industrial activity. It was praised by the economic pundits as a powerhouse of national wealth. And it was burnt into the national consciousness by a series of artistic images from Charles Dickens's *Hard Times* (set in 'Coketown', which was an evoca-tion of Preston in the 1850s) to Lowry's sooty scenes of Salford in the 1930s. It was, in short, the landscape of King Cotton, the achievement of which must be regarded as one of the most sensational and enduring features of the British Industrial Revolution.

For all this, the traditional textbooks have been misguided in identifying the Industrial Revolution with the transformation of the cotton industry. Important although it undoubtedly was, the cotton industry was only part of an all-embracing process of change in Western Civilization. We have already noted its effects on other key productive industries, and we will have more to say about its social implications, but to concentrate attention on the cotton industry to the comparative neglect of all these other factors is to get our understanding of the whole process out of focus. In particular, it tends to give a disjointed and episodic character to the process of industrialization, by suggesting that it had a sudden beginning, with the spinning inventions of the mid eighteenth century, and a definite ending, when the transformation of the cotton industry was complete, about 1830. In fact, of course, the 'beginning' took place in an environment already deeply influenced by changes in agriculture, coalmining and heavy industry, and the 'ending' occurred at a time when the impact of the railways was only just beginning to be felt, and when the mechanical properties of electrical power were in the process of being discovered. It is, therefore, unrealistic to take the cotton industry on its own as in some way dictating the chronology of the Industrial Revolution. It makes more historical sense to see industrialization as a continuing process of transformation, in which the cotton industry provided one amongst many instances of spectacular change.

Another point which is worth making in order to put the British cotton industry into better focus is that the transformation of the industry interacted with developments in the other textile industries. The woollen-cloth industry was much older and larger than the cotton industry in the eighteenth century, with a correspondingly greater investment in buildings, equipment and a skilled labour force. These factors were of great value to the cotton industry in its early days, as they meant that much of the infrastructure necessary for the development of a new industry was securely in place. They also meant, however, that the inertia of tradition and established practices made it more difficult to innovate in the woollen industry, but under the stimulus of rapid change in the cotton industry, the older industry was prompted to undergo a similar series of innovations, in many instances taking over machines such as the power loom, which had originally been designed for the cotton industry. As a result, the woollen-cloth industry, in both its woollen and worsted branches (the latter used longer fibres of wool, which required different preparatory treatment from that of the short-staple fibre used in woollen cloth, and produced a more durable fabric) underwent a process of mechanization and of concentration in large factories which was only less obvious than that in the cotton industry because it took place in areas such as the West Riding of Yorkshire in which the industry had already been long established. Similarly, although on a smaller scale, the silk and linen industries acquired the characteristics of mechanized factory industries in this period. Indeed, the silk industry has a valid claim to be the first factory-based industry in Britain, because the brothers John and Thomas Lombe built a water-powered mill on the River Derwent in Derby in 1717. The mill was 500 feet long and five or six storeys high, and can be regarded as the first factory designed to accommodate a series of power-driven machines. The machines in this case were silk-throwing machines, the secret of which the Lombes had acquired from Italy, and the need to protect their innovation was a major motive in building the factory. In the case of the linen industry, similar large mills were subsequently built to house the 'beetling' machines which were used to beat the finished fabric under a series of hammers in order to produce a lustre.

If there is any one image with which the new textile industries were associated, therefore, it is the factory, in the sense of a large and usually multi-storey building housing power-driven machines. There had already been examples of large-scale productive industries, such as some of the mining operations and ironworks which we have considered. Matthew Boulton had built up a large establishment of engineers and craftsmen at his Soho works in Birmingham, ideally equipped to manufacture Watt's steam engine. William Champion ran an enormous brass works in the mid eighteenth century at Warmley, near Bristol, which employed over a thousand people, and several of the naval dockyards of this period probably employed even more. But the factory was something different: it represented mechanization rather than handcraftsmanship, and water or steam power rather than manual power. Above all, it represented a regulated, disciplined environment, controlled by the clock and run like clockwork. To those who disliked it or feared its consequences, it was the 'dark Satanic' mill. To the entrepreneurs who invested in it, the factory was a guarantee that they could supervise their expensive equipment and maximize its use. To those, the new labour force, who worked in it, the factory provided a livelihood for which they had good cause to be thankful, but it also imposed a discipline of unremitting drudgery and appalling degradation. To all, however, the factory represented a means of wealth creation, and as such it was destined to stay as a feature of modern industrialization. So the emergence of the factory system was a pivotal aspect of the processes of modern industry.

In this treatment of industrialization down to 1850 the emphasis has been on Britain, and there are good reasons for this. The eighteenth and nineteenth centuries were, after all, the period in which Britain led the world in the development of industrial processes: whatever we mean by the expression 'Industrial Revolution', it certainly acquired its most dramatic characteristics in Great Britain in this period. The causes of this British primacy were largely fortuitous: it is probable that industrialization would have happened somewhere in the West, and if it had not been for adverse political and social circumstances it would probably have happened in France, then the richest country in Europe and the acknowledged cultural leader. But French industry laboured under the restrictions of an archaic monarchical system until these were thrown off in the French Revolution, and the Revolution itself set back French industrialization for two decades. So Britain, with its comparatively open society and freer system of trade and industry, provided the necessary stimuli to enterprise and the rewards for successful innovation which promoted the massive acceleration in the processes of industrialization with which we have been dealing.

The other countries of Europe and North America hurried to catch up when the Napoleonic wars ended in 1815, but by then Britain had acquired a substantial lead, and it was only after 1850 that it began to feel the competition of its rivals. Up to 1850, Britain was by general consent 'the workshop of the world', and the Great Exhibition of 1851 was a well-justified celebration of this dominant role in the Industrial Revolution. The combination of developments in coal, iron, steam and cotton had moulded a novel landscape in which the factory was the most striking feature. It demonstrated a new type of large-scale, coordinated industrial activity with an astonishing capacity for creating wealth.

6. The Age of Mass Production

The technological transformation of the processes of production continued to gather momentum in the second half of the nineteenth century, but its focus in industry and in location changed profoundly. The main stimuli to expansion were steadily growing markets for all sorts of raw materials and manufactured commodities, which promoted ever more complex techniques of mass production, and the availability of electricity as a source of power, which caused a rapid dispersal of manufacturing from the traditional concentrations of production. The increase in the scale of production generated significant structural changes within the manufacturing process, as the need for efficient control of complicated organizations encouraged the development of 'scientific management', and the corresponding need to keep ahead of actual or potential competitors caused these same organizations to invest increasing resources in the processes of research and development.

Whereas the expansion of production before 1850 had been characterized by the appearance of the factory system, the continued expansion thereafter has thus been marked more prominently by organizational adjustments to the needs of mass production. This does not mean that factory development was not important after 1850, nor that mass production was not present before that date. Large factories are still being built, although the change from steam power to electricity has brought about a spectacular dispersal of industry, coupled with an increasing use of road vehicles for transport. Modern factories are, moreover, far cleaner than their predecessors, which had spewed smoke and other waste products over the adjacent landscape without constraints of any sort. They also tend to be more temporary structures – sheds erected for the efficient performance of particular processes and the comfort of their workers, and capable of rapid modification or replacement – and so their overall impact on the landscape is much reduced. It would be true to say, however, that the factory remains a basic unit of modern productive industry, and that its essential form has been adopted in other processes, such as farming.

As far as mass production is concerned, this also shows more continuity than a simplistic dichotomy of before and after 1850 would suggest. To go no further back than the beginning of the nineteenth century, the set of machines designed by Sir Marc Brunel (father of I. K. Brunel) and made in the workshops of Henry Maudslay to produce the large quantities of wooden blocks required for the rigging of sailing ships in the British navy was a clear anticipation of mass-production techniques. It involved over forty machines, each performing a specialized function in shaping or assembling the parts of a single block, and the effect of these working in combination was enormously to increase the output of manufactured blocks at a considerable saving in the amount of labour required. Virtually all the ingredients of a modern mass-production process were present, except for a moving assembly line to make the interconnection between one machine and another.

By the middle of the century, the idea had been taken up and elaborated. At the Great Exhibition of 1851, which in so many respects represented the high-water mark of British achievement as the 'workshop of the world' in productive industry, the assembly of the Crystal Palace in Hyde Park in an astonishingly short time was made possible by casting standardized iron parts and producing unprecedented quantities of plate glass. But it was in the Exhibition itself that the outstanding marvels of mass production were to be found, especially in the small arms and agricultural machinery made by American manufacturers for the domestic market as it expanded into the interior of the continent. The manufacture of revolvers, rifles, harvesters and threshing machines became within a few decades such a characteristic feature of American industry that it was known, somewhat misleadingly, as the 'American System', with standardized parts manufactured on the principle of complete interchangeability, so that any broken or defective part could easily be replaced. The same system was soon adopted for the manufacture of sewing-machines, typewriters and other pieces of mechanical equipment for home, farm and office. Even though the United States of America had no monopoly of the system, there is no doubt that it led the world in exploiting the advantages of systematic mass production, and thereby gained the initiative from Britain and other European nations in the processes of industrialization. It was, therefore, a development of momentous importance in the history of technology.

Mass production of machine parts only became possible with the development of machine tools which could produce the parts required to the necessary standards of accuracy and reliability. Traditionally, the skills of the engineering craftsman depended upon his aptitude, training and personal tools, and there are many testimonies to the excellence of these skills amongst the pioneers of industrialization in Europe and America. Given time and the necessary resources, such craftsmen were able to fabricate complicated machines such as the early steam engines and locomotives, and to build mills and factories. But their productivity was severely limited, and could only be increased by mechanizing some of their skills; in other words, part of the skill in using the craftsmen's tools had to be transferred to machines, and this required the construction of better machines for making machine parts than any which had previously been available. The same Henry Maudslay who had assisted in the manufacture of Brunel's block-making machinery also pioneered this role. By insisting that all the machine tools in his workshops should be made of metal, and by introducing such refinements as the slide-rest to the lathe, the basic metal-working machine tool, Maudslay achieved new levels of excellence in replicating precise mechanical specifications. He died in 1831, but by that time his workshop in Lambeth had served as the training ground for a generation of mechanical engineers, among them Roberts, Nasmyth and Whitworth, who carried on the Maudslay methods with great success, completing the transformation of what had been essentially a craft practice into the modern machine-tool industry.[1]

When American manufacturers began to develop the potential of fully interchangeable machine parts the infrastructure of mass-production industry in the shape of a range of versatile and precise machine tools – lathes, drilling machines, milling machines, planing machines and machines for performing a host of other specialized operations – was thus firmly in place. What they did was to refine many of the machines, as, for instance, with the introduction of a turret providing a choice of cutting tools on what became known as a 'turret lathe', and to apply them on a large scale to the systematic manufacture of specialized parts which could be readily assembled into the final product. The next major step in the evolution of mass-production systems came with the introduction of assembly-line techniques, exemplified in the huge automobile-manufacturing plant established by Henry Ford at Detroit at the beginning of the twentieth century. Such techniques are only appropriate for high-volume consumer goods such as motor cars, and are less applicable, say, to shipbuilding or even to the manufacture of aeroplanes. But the thinking behind such massive productive establishments involved something of a quantum jump in the philosophy of industrial organization. It was closely associated with the emergence of scientific management, and as such permeated twentieth-century conceptions of productive industry, with implications for all aspects of technological revolution. It will be necessary to consider these implications in more detail, but before doing so it will be useful to review other aspects of productive industry in the century and a half since 1850.

We have observed the strategic part played by the bulk production of iron, in industry and in society, before 1850. Up until that time, the iron-carbon alloy, steel, albeit important in precision instruments and in tools requiring hard-wearing cutting edges, had remained an expensive substitute for iron, being available in comparatively small quantities through batch production in the crucible or cementation processes. A transformation of the heavy iron and steel industry was initiated in 1856 by Henry Bessemer with his invention of the converter, a method of decarburizing molten cast iron by blowing a blast of air through it, which had the effect of producing a form of mild steel in bulk for the first time. The process encountered severe problems in its early years, particularly with its failure to remove unwanted elements from the iron in the converter, but remedies were eventually found for these difficulties, and entrepreneurs who were stimulated to search for alternative methods of mass-producing steel devised the very successful open-hearth process, which could be controlled more precisely than that of Bessemer and which could easily be adapted for scrap iron. By the end of the century both processes were in widespread use throughout the iron and steel industries of Europe and North America, and steel had replaced iron for most uses in machines, and also for many constructional purposes, in ships, buildings and railway lines. The British iron and steel industry had led the way in this transformation, but manufacturers in France, Belgium, Germany and the United States had been quick to follow, and by the beginning of the twentieth century Germany, with its massive concentration of heavy industry in the Ruhr, and America, with the great developments around Pittsburgh and elsewhere, had overtaken Britain in steel production. Steel, indeed, had become the pervasive and dominant metal of the Western world.

Nevertheless, other metals have become more rather than less important in the economy of the rapidly industrializing societies of the twentieth century. Copper has acquired great significance in electrical engineering, on account of its high conductivity, and tin is used in food preservation – thin sheets of steel are covered with a veneer of tin which can then be rolled and sealed to form the ubiquitous tin can. With the discovery and exploitation of easily available deposits of both copper (in America) and tin (in Malaysia), much of which could be mined by open-cast techniques, the traditional deep mines of Cornwall and Devon declined in importance because the costs of extraction were so much greater. Deep metal mining has remained important, however, especially in the mining of gold in South Africa and elsewhere. World demand for gold, as a basis for national currencies, has continued to grow, promoting a series of 'gold rushes' in California, Australia and South Africa, and this has ensured the continuity of the techniques worked out in Cornwall and Devon, but with further refinements such as steel rope for winding, pneumatic drilling and electrically powered winding engines and ventilating machines.

New metals have also been exploited, most notably aluminium. Although occurring widely, this could only be extracted from its ores with great difficulty before the discovery in 1886, in both France and the United States, of an electrolytic method of producing the metal in bulk and comparatively cheaply. As this process consumed large quantities of electricity, the major aluminium smelters tended to develop close to abundant supplies of hydroelectric power, such as the French Alps and the Highlands of Scotland. Aluminium has come to be widely used for mechanical and structural purposes, often as a substitute for steel in circumstances where the comparative lightness of aluminium is an advantage, as in aircraft frames and many kinds of engine.

While the extraction and utilization of metals has continued to expand in the twentieth century, causing serious apprehension about the eventual exhaustion of available deposits in the foreseeable future, the extraction of coal has diminished. As we have seen, this played a crucial role as the fuel for steam engines in the early stages of rapid industrialization, and it reached an all-time peak, in Britain and other Western countries, at about the time of the First World War. Since then, however, the decline of the steam engine and the corresponding rise of the internal combustion engine, with the consequential massive increase in fuel-oil production, has brought a contraction of the market for coal and a reduction in extraction. Coalmining has continued to be an important industry, and, like metal mining, it has adopted electrification and other new techniques, but it has become a smaller and altogether more compact industry than it was eighty years ago. This has reduced the pressure on coal reserves, at least so far as the developed countries are concerned, and has encouraged a concentration of mining in the more productive coalfields and a clearing up of some of the physical and social debris of traditional coalmining areas. Mining communities now tend to be smaller and politically less potent than they were earlier in the century, but they are also less isolated and better integrated into the general life of the area.

Coal came to play an important part in the chemical industry in the second half of the nineteenth century, not only as a fuel but as a subject for chemical analysis. The chemical industry had grown impressively since the middle of the eighteenth century, both in such traditional forms as ceramics and glass manufacture, and in the new heavy chemical industry producing acids and alkalis in bulk for industrial use. The scientific examination of coal tar, acquired as a waste product from town gasworks, instigated an entirely novel development – the creation of an organic chemical industry producing a wide range of materials which would transform many aspects of life and society. The interest of Prince Albert, the Prince Consort, had secured the appointment of a fellow German, A. W. von Hofmann, as principal of the Royal College of Chemistry in the 1840s. Hofmann had been a student of the great German chemist von Liebig, and he carried on his mentor's study of the derivatives which could be obtained from coal tar in his new office, setting his students to work on what came to be regarded as a new branch of chemistry – the chemistry of carbon compounds, or organic chemistry. One of these students was W. H. Perkin, who had the distinction of making the first significant application of organic chemistry by producing a synthetic dye. This was in 1857 when, after trying – and failing – to produce quinine from a coal-tar-derived aniline compound, he discovered accidentally that the material he had isolated was a very effective mauve dye, which he quickly went into business to manufacture. This was the first of a series of synthetic dyes which largely replaced natural dyes in the British textile industry.[2]

After this initial success in Britain, most of the development of the synthetic dye industry took place in Germany; organic chemistry made a significant contribution to the rise of this new nation state as the dominant industrial country in Europe, surpassing Britain in several respects by the end of the nineteenth century. Analysis of another familiar material – cellulose from wood pulp or other vegetable matter – led in the same period to three other outstanding developments in the organic chemical industry: high explosives, artificial textiles and plastics. High explosives, from compounds of nitric acid and cellulosic material, transformed warfare and justified the description of the First World War as 'a chemists' war'. Artificial textile fibres were produced by extruding a cellulosic mixture into various hardening agents. Very delicate filaments could be made in this way and used as artificial silk, or rayon; by the end of the nineteenth century it was being marketed to the textile industry, usually for mixing with natural fibres. Later came nylon and the related family of artificial fibres and these new materials were recognized as being stronger and more versatile than natural fibres. Meanwhile, the first of the new artificial plastics – to distinguish them from natural plastics like rubber, which also underwent a big industrial development in the second half of the nineteenth century – were being made from cellulosic mixtures, by adding hardeners; 'celluloid' was used for billiard balls, gentlemen's collars and photographic film by the end of the century. In 1907 the Belgian chemist L. H. Baekeland produced a gummy material by mixing formaldehyde and phenol; when heated, it could be moulded to set as the hard material marketed as Bakelite. Its excellent insulating qualities led to its widespread use in electrical fittings, and most early radio sets had a moulded Bakelite case designed to attract the domestic customer.

The organic chemical industry began with the investigation of coal-tar derivatives, but in the twentieth century it has come to rely heavily on the products of the oil industry. This barely existed in 1850. Bitumen had been known from antiquity as a source of oil for providing light, but it was not until 1859, when the first successful bore to tap subterranean sources of crude oil was made in the United States, that an oil industry could begin to emerge. Its development was delayed by the American Civil War, and it was only after that disruption that systematic attempts were made to extract the newly available material in bulk. A large industry then quickly grew, with pipelines, simple distilling apparatus to extract the required fractions (the portions of the crude oil that vaporize between specific temperatures) and distributors supplying barrels to convey the product to the customers. These were mainly people living beyond the reach of an urban gasworks who needed some illumination at night and who welcomed the new kerosene as an alternative to the traditional candle or taper.

Kerosene (paraffin in British usage) came from the middle fraction of the distilled oil, and for some time the most volatile part, gasolene (British petrol), was regarded as a potentially dangerous waste product. The adoption of the internal combustion engine to power vehicles changed all that, because the very qualities which made gasolene so dangerous – its lightness, volatility and easy combustibility – were ideal for use in a locomotive engine. There thus dawned in the 1880s an era of prodigious growth for the oil industry with the universal popularity of the automobile and later the aeroplane, as well as many other applications of oil-fuelled engines and boilers which developed alongside them. A worldwide search was made for likely underground reserves of oil, and giant international enterprises grew up to exploit these sources and to transform their products into a widening range of usable materials. The industry had profound political and social ramifications, as it created international tension in the Middle East and brought instant affluence to some desperately impoverished parts of the world. It also involved an enormous traffic in bulk oil, mainly by tankers at sea when overland pipelines were not feasible, and spectacularly large refineries sprang up to perform the chemical distillation and to produce the required fractions. In addition to the indispensable fuel oils, oil produces derivatives which became essential to the manufacture of plastics, artificial fibres, insecticides, fertilizers and pharmaceutics. The dependence of the major industries which produce these commodities on oil-based materials has increased the dominance of the petrochemical industry and its complex of refineries which 'crack' the crude oil and undertake complicated processes of molecular engineering to prepare it for their customers.

The only range of industries which can compare in importance with the petrochemical industries in the period since 1850 is that associated with the application of electricity. This is not an easy comparison to make, as the impact of electricity has been more general than that of the chemical industry. In one sense everybody has been affected, because the adoption of electricity for lighting and most small-scale power uses has been virtually universal in modern Western societies. But there has also been a large and flourishing engineering industry associated with the generation of electricity, and with the construction of electric engines, locomotives and other equipment. Even more momentous, however, has been the proliferation of the electronics industry, which began in a modest way with the production of radios and thermionic tubes ('valves' in British usage) in the 1920s but which has expanded enormously with the demand for television sets, computers and calculating machines which took off after the Second World War. Every Western country now has its equivalent of California's 'Silicon Valley' – an area of many modern factories, mostly of medium size, all employing highly skilled labour on the manufacture of sophisticated electronic apparatus or preparing elaborate computer programs to be marketed as software. The location of these industries has usually been determined by non-economic factors such as an attractive climate or the proximity of a university town, although the provision of a good transport system and of other desirable public services may be taken for granted.

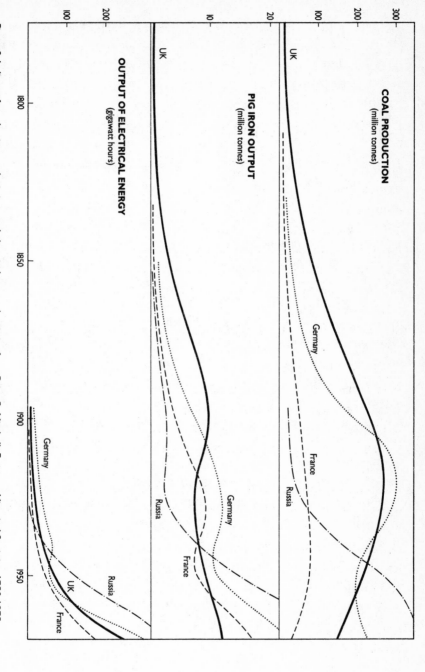

Figure 8. Indices of production: coal, pig iron and electrical energy (statistics from Brian R. Mitchell, *European Historical Statistics, 1750-1975*, 2nd edn, London, Macmillan, 1980)

These complexes of electronics-based enterprises, grouped on the motorway corridors or on the fringes of our university cities, have become one of the most characteristic developments of late-twentieth-century industry and society. Compared with the 'dark Satanic mills' of the early phases of industrialization, they are wholesome and desirable places in which to live and work. They probably represent the sort of industrial landscape towards which advanced industrial societies are tending, especially as a growing concern for the environment and for the avoidance of pollution has put even old-established industries under pressure to improve their performance and their public image. But many of these older industries remain important – iron and steel, metal-working, coalmining, engineering and heavy chemicals are still indispensable to production in the twentieth century and, as we have seen, many of them have made striking advances in this period. It is difficult, however, to soften the environmental impact of, say, quarrying or metal extraction, and even the petrochemical industry has problems in this respect as it handles some materials of extreme toxicity, so that accidents, when they happen, tend to be distressingly serious. Despite these problems, recent decades have witnessed an overall improvement in environmental standards, even though there is clearly no cause for complacency and industry is only just beginning to come to terms with acid rain and the greenhouse effect. And it is a disturbing commentary on the insularity of modern attitudes that many of the industrial abuses which have been remedied in the West are now being replicated in the developing countries.

The conjunction of many of the factors which comprise the modern industrial scene and pose serious environmental problems can be seen in the most ancient of all industries – that concerned with the production of food and drink. Farming in the advanced countries has been subject to the same processes of rationalization which have characterized other large organizations in the twentieth century. Labour-saving devices from tractors to electric milking machines have been generally adopted, while fields have been made larger to accommodate combine harvesters and other machines more efficiently than the traditional small fields of peasant or tenant farmers. The land has been purged with insecticides and enriched with artificial fertilizers, both bought at considerable expense from the chemical industry, and both causing a pollution problem affecting wildlife and water supplies. Factory methods have been deliberately introduced into stock farming, with battery hens, pig houses and intensive beef-rearing units. While the labour force has been reduced on modern farms, the costs and organizational complexities of the new regime have proved too complicated for some old-fashioned farmers and have created pressures for large-scale management control. The productivity of modern farms has undoubtedly increased, and it is arguable that such methods are necessary in order to ensure the supply of food and drink without which our quality of life could not be maintained. But there can be no doubt either that this has been achieved at a cost, both to the landscape and to the intimacy of man's relationship with the natural world, and involves serious pollution hazards.

Here, as elsewhere, technological revolution appears different according to the perspective from which one views it. From the point of view of the vast majority of the world's population, the high productivity of farming in Western societies must seem extraordinarily luxurious, and until serious steps are taken to close the gap in relative living standards it is not appropriate either to consider overproduction a problem in the Western farming industry or to criticize desperate (and probably misguided) attempts to increase the productivity of farming in the developing nations by overgrazing the land or by drastic forest clearance. The farming industry, in short, is part of a world problem, which requires a world solution.

The facilities for moving food over long distances have been increasingly available since 1850, but so far have been used almost entirely to transfer food produced on the periphery of the world community to those parts of Europe and North America which have instigated and controlled the exercise. Bulk transport by railroad and steamship brought American grain to Europe in the second half of the nineteenth century. Refrigeration techniques, developed effectively in the 1870s, were used in conjunction with reliable steamship services to bring meat, dairy produce and fresh fruit to Europe from Argentina and the Antipodes. Efficient canning procedures were developed in the mid nineteenth century to extend the market for American beef and other foodstuffs. More recently, freeze-drying and irradiation have proved to be useful ways of preserving food, and they are likely to be extensively adopted in the Western world.

As with most other aspects of productive industry since 1850, the quantitative increase in food and drink production has been accompanied by a corresponding qualitative change in the nature of the organization required to fulfil the needs of the expanding market. In short, there has been a transformation of management structure which deserves to be considered as part of the process of industrialization in the modern world. The genesis of modern management goes back long before 1850. Substantial eighteenth-century industrial organizations such as Wedgwood's Staffordshire pottery or the Soho Foundry in Birmingham, owned and managed by Matthew Boulton, gave plenty of attention to organizational efficiency, the elimination of waste and the well-being of their workers, and they anticipated in some respects the development of managerial skills in later enterprises. But like all such organizations in Britain – with the exception of government-sponsored ventures like the naval dockyards, which were large but tended to be dominated still by the traditional attitudes of administrators and craftsmen – these were private companies under the direct and continuing control of their owners.

The development of the large statutory companies, starting with the canals and undergoing a huge expansion with the advent of the railways, and then the general provision for the limitation of liability in public companies, introduced in Britain in the 1850s, changed the scale of operations and meant that it was now necessary to bring in managers. Managers did not own the concern which they controlled in the interests of the shareholders, even though they might have a share in it themselves. They were employed as salaried officers to perform specialized jobs, and could be dismissed by the owners, be they the shareholders of a public company or the government (in the case of a dockyard or military establishment), without prejudice to the integrity of the undertaking. Such managers gradually became more important in British industrial practice during the second half of the nineteenth century, even though owner-managers like John Brown in steel-making and William Lever in the manufacture of soap continued to dominate the scene. With the inexorable increase in the size of the leading enterprises, however, such individual and family links became steadily weaker in the twentieth century and, so far as large companies were concerned, the managers came into their own. A similar development took place in continental Europe and North America, beginning rather later than in Britain but then proceeding faster so that, by the beginning of the twentieth century, large enterprises like that of the Krupps family in German steel-making, or Henry Ford in American automobile manufacture, were becoming exceptional amongst huge managerially controlled corporations, frequently formed by the combination of many smaller units. And, ironically, even Krupps and Ford adopted many of the managerial strategies in order to maintain their spectacular expansion.

The social impact of managerialism, in respect of such features as professionalism and training in management skills, will be dealt with elsewhere, but here it is desirable to note the profound importance of managerial organization on the pattern of development of modern productive industry. Managerial thinking began to generate a 'philosophy' of industry at the beginning of the twentieth century with the work of the American engineer Frederick W. Taylor, who published his treatise *Scientific Management* in 1906. This was devoted largely to what became known as 'time and motion study' – the skill of organizing the workforce in such a way that it ensured the most efficient and therefore the most profitable production. This method achieved impressive successes and it was enthusiastically adopted by many American enterprises, most notably by Henry Ford, who devised his assembly-line organization on strict work-study principles. It is not surprising to find that not only were Taylor and Ford personal friends, but that they were both friends of Thomas A. Edison, the pioneering inventor whose research laboratories had anticipated some of Taylor's recommended procedures. American managerialism was grafted easily and comfortably on to the older tradition of American individualism.

While Taylor's methods were widely taken up by managerial organizations in the first two decades of the twentieth century, it soon became apparent that they did not solve all the problems of industrial production as thoroughly as their early exponents had maintained. In particular, it was recognized that there was a certain lack of sophistication about methods of work study which saw the workers only as units of production: they generated resistance on the part of the labour force and beyond a certain point no further application was beneficial. The human element in the managerial equation was rediscovered by another American investigator, Elton Mayo, in what became known as the 'Hawthorne Experiment' after the name of the American industrial plant where it was conducted. Mayo came to the seemingly astonishing conclusion that people worked best when they were regarded as responsible persons, to be consulted and involved in the process of improving industrial efficiency. This made a big impression on managerial thinking in the 1920s and 1930s, and the new approach was largely adopted in the corporate strategy of the more advanced companies.

Both the principles of work study and the concern to involve workers in the success of the enterprise have been subsumed in more recent developments in managerial practice, which have been animated by a form of structuralism in the sense that they have tried to regard the organization as a whole and to devise the most efficient 'flow' of functions and responsibilities throughout the structure. In current managerial philosophy it is thought important to devise a management plan and to apply performance indicators at every level of operation, with procedures for dealing rapidly with any problem and for monitoring the attainment or otherwise of production targets. Management, in the last resort, depends on the qualities of the people practising it, and it can never be perfect, but it would not be unreasonable to maintain that modern management techniques, as they operate in the industries of the Western world, have achieved a remarkably high level of overall efficiency. They have demonstrated very impressively that large organizations can be operated as efficiently and humanely as any traditional small company, and have set the dominant industrial pattern in which all modern industrial production takes place.

There is one aspect of the pattern which is worth special attention as a postscript to this discussion of the development of corporate management. This is the recognition of the continuing need for study of the product of the organization, or what has become known as 'research and development'. Any farsighted entrepreneur of a traditional small company would have shared this perception, and we have mentioned some, such as Josiah Wedgwood and Matthew Boulton, who took an active interest in it. But few of them had the resources to undertake anything on a substantial scale, and it was not until the late nineteenth century that enterprises began to make a systematic attempt to provide facilities for research laboratories and to employ the right sort of scientific experts to pursue this objective. Again, it was Thomas A. Edison who demonstrated the effectiveness of a research organization pursuing systematic and persistent research into product development. Edison became known as 'the Wizard of Menlo Park' after the research laboratories which he established in New Jersey.[3]

The electrical and electronics enterprise built upon Edison's exploitation of the electric light bulb and the phonograph – General Electric – has been amongst the foremost American firms to establish large research laboratories. But other American corporations were quick to follow suit, like the American Telephone and Telecommunications company, A T and T, which was formed to exploit Bell's invention of the telephone and developed the formidable Bell Laboratories in which the transistor was invented in 1948. Even in Britain, where it has long been something of an academic cliché to chastise industrialists for their lack of interest in research, some firms have had a distinguished record. The work of Robert Hadfield's and other Sheffield steel companies in studying and exploiting new steel alloys has recently received approving academic attention,[4] and initiative was also being shown by British chemical companies from the beginning of the twentieth century.

The fact is that an effective level of research and development is now recognized throughout the world of big industrial enterprises as an indispensable element of corporate achievement. It has been assimilated into the managerial objectives of modern organizations for industrial production, thereby doing much to maintain the dynamics of mass production and industrial growth.

7. Transport before the Railway Age

A good transport system is an essential feature of a developed industrial society, making possible the easy movement of goods and people at reasonable cost and with maximum convenience. It is hardly surprising, therefore, that the emergence of such a worldwide transport system has been a prominent ingredient in the transformation wrought by technological revolution in the last three centuries. One improvement has followed another in a spectacular series of technical achievements, which have made possible the movement of vast quantities of raw materials and manufactured goods all over the world and have reduced travelling times between Europe and the Americas or the Antipodes to a few hours. In this process of rapid technological change, the 'package' and 'ratchet' mechanisms have repeatedly operated. Each development has occurred where – and only where – the necessary combination of needs, resources and enterprise have been present, and where sociopolitical conditions have been congenial to their occurrence. Time after time, moreover, a development has served to prepare the way for new improvements which have replaced the original development or made it redundant. And in all this complex process of change, the human and contingent factors have remained irreducible: the strategic priorities of ambitious rulers and the perceptions of brilliant engineers have made their mark, while some ingenious and technically viable schemes, such as the idea of an 'atmospheric railway', with trains propelled by a partial vacuum in a tube between the rails, have been abandoned as commercial disasters. In this chapter we will review the transport revolution to the middle of the nineteenth century, and we will pursue the theme down to the present in the following chapter.

It is difficult to imagine the immobility of most people before the advent of modern transport systems, particularly as traditional historical accounts tend to emphasize the activities of the few – merchants, missionaries, soldiers, sailors, political adventurers and such like – who did move comparatively freely. But the great majority, even in societies not dominated by serfdom and other restrictions such as poverty and lives of unremitting drudgery which kept people tied to their place of work, rarely moved far beyond the village community into which they were born. There might be occasional visits to the local market town or fair, a rare expedition to a regional capital, and the outside possibility of a once-in-a-lifetime pilgrimage to a national cultural shrine. But for the most part, life in Western Europe was almost immobile once the great population movements of the Teutonic and Viking invasions had stabilized around AD 1000. As late as the eighteenth century, people tended to remain settled, if only because of the physical difficulty of moving in the absence of reliable means of transport. The series of dramatic innovations by which this situation has been transformed in little over two hundred years constitutes a significant part of the process of industrialization and technological revolution.

It was inland transport which was the main problem; people living close to the sea and navigable rivers fared rather better. Inland transport depended on animals and roads, and whereas there was a good supply of horses, at least for the well-to-do members of the community, roads had been neglected ever since the days of the Roman Empire. The reasons for this neglect were political and social: the states of medieval and early modern

Europe lacked the resources and any overriding military imperative to provide a viable network of roads, and responsibility for
maintaining the highways was frequently farmed out to local
magnates and village communities who had neither the interest
nor the expertise to build good roads. The result was that most
European roads were, at best, patchy, with only sporadic attempts
to provide proper foundations and a paved surface, and more
usually consisted only of tracks, which became rutted and worn
and, in bad weather, impassable. Consequently, any form of
vehicular traffic was difficult to maintain. It had to be ponderous,
like a farm cart, in order to be sufficiently robust to survive the
rough surfaces, and that meant it had to be slow, and also that it
caused further damage to the existing road surfaces. Thus most
traffic went by horseback, goods being carried in panniers slung
over the withers of the horses, which made the transport of bulky
commodities prohibitively expensive. Carriages, such as they were,
were the prerogative of royalty and the nobility, and even these
proceeded slowly and frequently relied upon the assistance of
additional teams of horses or oxen to get them over the worst
stretches of road.

The contrast between difficulties of transport in the land-locked
interiors of Europe and the comparative mobility of citizens living
by the sea or on large rivers was marked. Boats emerged very early
as a technological innovation amongst riverine communities; as
they developed they provided an effective means of transport
for people and commodities wherever the natural watercourses
permitted. At sea, boats tended to be larger – the sailing ship had
been known from antiquity. Improvements over the centuries in
ship construction and the increasing mastery of sailing techniques
placed at the disposal of Western societies a remarkably versatile
form of maritime transport, which required no fuel after the initial
fitting out and which became capable of carrying a substantial

cargo over the seas of the world. Sailing ships provided much of the vital mercantile wealth without which the subsequent processes of industrialization could never have started. But they made no direct contribution to the solution of the problem of inland transport except, perhaps, to increase the frustration of inland producers who could not get their goods to market and thus to stimulate them in the search for improvements.

The first such improvements came in the form of artificial waterways. Canals for drainage and irrigation purposes had been known in the ancient world, but in Europe the uses for such waterways were limited in the Middle Ages and the expertise in building them languished. Interest in artificial waterways revived in the sixteenth and seventeenth centuries, with the thriving towns of northern Italy, Flanders and the Netherlands beginning to construct short stretches in order to improve the trade in their products. The first really modern canal, crossing a watershed with the aid of pound locks, was the Canal du Midi, completed in France in 1692, to connect the Atlantic with the Mediterranean. It was conceived as part of the grand military strategy of Louis XIV, so that the French fleet could be deployed more easily, but in fact its main function came to be commercial, even though the region of southern France through which it ran was not one of great industrial activity.

Continental canal-building provided an inspiration to British canal construction; the young Duke of Bridgewater visited canals in France and Italy and was encouraged to introduce them on his own estates in Lancashire. His first objective was to get coal from his collieries at Worsley, a few miles north of Manchester, into the middle of that rapidly developing town, by what became known as the Bridgewater Canal. This was completed in 1761 and the duke subsequently promoted its extension to the navigable Mersey at Runcorn, and then the link between the Mersey and the River Trent, which became the Trent and Mersey Canal. The Bridgewater Canal was not the first artificial waterway in Britain, as there had already been significant river-improvement works on the major rivers of the country, and the Exeter Canal had given sea-going ships access to that city from the sixteenth century. Moreover, both the Newry Canal in Northern Ireland and the Sankey Brook Canal in Lancashire had already incorporated pound locks. But the Duke of Bridgewater deserves credit for the flair and persistence which he brought to pursuing the commercial potential of canals.

The idea of commercial canals was taken up with enthusiasm, especially by entrepreneurs producing goods in inland areas such as the Staffordshire Potteries and Birmingham, and within a few decades Britain was equipped with an extensive network of canals linking all the main river systems and providing access to the sea for most of the land-locked areas of the country which possessed the resources for the development of manufacturing industries. Coal and iron, china clay and manufactured pottery, sand and glassware, grain and stone, bales of cotton and wool, were amongst the bulky cargoes which could be carried advantageously on this new network, thereby adding significantly to the overall economic activity of the country and contributing to the rapid acceleration of the processes of industrialization in Britain.

The early canals in Britain were utility affairs. Once the route had been determined by Act of Parliament, the device of a statutory company made possible the compulsory purchase of land, but it was difficult to raise the large amounts of capital required. This meant that the canals were built with the minimum of frills and an emphasis on economy in engineering. Most of them were 'narrow' canals, with locks seven feet wide, which gave rise to a distinctive type of narrow boat capable of using them. They tended to be sinuous, following the contours in order to avoid unnecessary changes in level, and using tunnels rather than deep cuttings when hill ridges had to be crossed. They were made watertight by tramping in loads of clay, a process known as 'puddling', and when river crossings could not be avoided, masonry or brick aqueducts were constructed to carry the puddled clay trench.

The men who built the canals were remarkable, not least because they contributed substantially to the creation of the engineering profession in Britain. The Duke of Bridgewater employed the millwright James Brindley to build his canals, and Brindley quickly demonstrated a natural genius for the job and became an engineering legend. He was a rough-and-ready sort of man, of humble origins and few social graces, but he discoursed with surprising eloquence to parliamentary committees concerning the virtues of canals and he showed his grasp of the engineering principles involved by constructing the first canal aqueduct and the first canal tunnel in Britain. Many followed him in becoming canal engineers, and a considerable proportion of them were trained on his canals. Most of them came from similarly humble origins, millwrights, stonemasons and other craftsmen, although a few had a professional background: Robert Mylne was an architect and John Smeaton came from a legal family. It was Smeaton who drew together this rather random group of men to form the first professional organization of engineers, the Society of Civil Engineers, which still exists as a dining club for senior British engineers. But it was the boom in canal construction which gave them steady employment and the opportunity to distinguish themselves as a new profession.[1]

The boom in British canal building peaked in the 1790s; by that time it had become easier to raise capital for canal ventures and so the second generation of canal builders, with men such as John Rennie and Thomas Telford, were able to build on a larger scale than their predecessors. They generally adopted a width of lock which was twice that of the traditional narrow lock; they built huge masonry aqueducts, and began to use cast iron for the trough which carried the canal over the masonry piers; and they undertook cutting and embanking in order to keep the line of their canals as straight as possible. These new canals sometimes replaced earlier narrow canals, but more usually they provided additions to the existing network, opening up further areas of the hinterland by the penetration of artificial waterways.

The engineers who emerged to build the British canals quickly found alternative employment in other transport undertakings, particularly in port improvements and road works. At the beginning of the eighteenth century most harbours were still essentially simple wharves on rivers and estuaries, and the growth of trade in ports such as London and Bristol was causing acute congestion of these facilities. In London, ships frequently had to be moored for many days in the river before they could be unloaded, and were prey to huge losses through pilfering. The solution to these and related problems came in the form of enclosed docks, providing increased wharfage at which ships could be safely moored in permanent high water, and where security could be provided by substantial warehouses and a high surrounding wall. John Rennie, William Jessop and Thomas Telford were among the engineers who built large enclosed docks alongside the River Thames at the turn of the eighteenth and nineteenth centuries. Liverpool already had such docks, built by the local engineers Thomas Steers and Henry Berry, and Bristol followed with its 'Floating Harbour' – a high-water dock created by damming the rivers in the centre of the city – constructed by William Jessop early in the nineteenth century. The major British ports were thus equipped to deal with the continued build-up of commercial traffic, with regular additions to the system of enclosed docks as they became necessary.

Meanwhile, other aspects of maritime traffic received attention from the new breed of engineers, with the systematic construction of sea walls and breakwaters, the dredging and straightening of harbour entrances, and the provision of lighthouses. John Smeaton had established his engineering reputation with the completion of the Eddystone Rock lighthouse, fifteen miles off Plymouth, in 1759. It was the first of a new type of lighthouse, the product of a coordinated team of consultants, masons, labourers and sailors, built entirely of interlocking masonry blocks cut to shape on the mainland and then assembled in their predetermined positions. Smeaton adopted the parabolic shape of the outer walls to deal effectively with pounding by waves, and in this as in many other respects his lighthouse set the pattern of all subsequent deep-water installations of this type. It was thus a very appropriate symbol for the Institution of Civil Engineers which, founded in 1818, received its Royal Charter in 1828 and incorporated Smeaton's masterpiece into its coat of arms.[2]

Engineers were also employed to improve British roads, but their contribution in this quarter was somewhat delayed by the difficulties already mentioned of raising sufficient resources for road building. As long as the parishes remained responsible for road construction and maintenance it was not possible to undertake any large roadworks; the first attempt to overcome this restriction by vesting the responsibility in special trusts which were allowed to charge tolls was at first only partially successful in achieving road improvement. The turnpike trusts, as they became known, were generally composed of local landowners and entrepreneurs who were reluctant to commit themselves to raising large sums of money, so that most of the early improvements were of a very modest nature. New roads required a surveyor, but these could often be hired from local estates, and any idea of employing a professional engineer was resisted. However, when substantial engineering work such as a large bridge became unavoidable, most trusts accepted the need to seek good professional guidance, and some of the surveyors also acquired considerable expertise in the basic techniques of road building. Called upon by the government to provide military roads in the Highlands of Scotland, for example, General Wade managed to train up army surveyors and road builders who did an impressive job in the middle decades of the eighteenth century, and elsewhere self-trained experts like 'Blind Jack' Metcalfe of Yorkshire succeeded in laying out the course of some sound roads.

Government resources were also available to Thomas Telford, both for the construction of the Holyhead Road from Shrewsbury and for his massive building programme of harbours, the Caledonian Canal, bridges and roads in Scotland. Telford's preferred method of road construction involved the use of heavy foundations which placed this type of improved road beyond the financial capacity of most turnpike trusts, and it came as a godsend to them when a Scottish surveyor, John Loudon McAdam, devised a simpler form of construction which dispensed with Telford's massive foundations, consisting essentially of an impervious skin over a carefully surveyed and well-drained under-course. This type of road had the additional advantage over Telford's that its surface, having some elasticity, wore better than the surface of a road with unyielding foundations. McAdam's expertise was quickly in demand from turnpike trusts all over the country. By the 1830s he and his family had been responsible for hundreds of miles of improved road.

It appeared, therefore, that the long-standing problems of inland transport had at last been overcome in Britain: a network of canals had been created which could carry heavy goods to virtually all parts of the kingdom and a corresponding network of viable roads facilitated the movement of people and lighter goods. Moreover, the opportunity for passenger transport offered by the improved roads had been seized by enterprises operating fast stagecoach services, and the design of these coaches had undergone rapid development since the mid eighteenth century. Dished wheels and good springing, with regular changes of horse teams, allowed them to maintain unprecedented speeds over long distances. They carried mail and passengers who could afford the fares to all parts of the country, and a variety of smaller carriages developed for other purposes in town and country. The construction of these coaches and carriages became a flourishing craft industry, and by the 1820s serious attention was being given to applying steam locomotion to the larger coaches. Several designs of horseless steam carriages appeared on the roads, and acquired a certain popularity even though occasional boiler explosions dented their public image. But it was other factors which came to threaten the traditional stagecoaches and canal traffic at this time; by the end of the 1830s both roads and canals were suffering severely from a new and highly competitive form of transport – the railway.

The roads eventually made an impressive comeback, thanks to the internal combustion engine, but this competition was almost fatal to the canals as far as their commercial prosperity was concerned. After the canal boom of the 1790s, construction was checked by the Napoleonic wars, and when it was resumed after the battle of Waterloo in 1815, it was on a much reduced scale. Some of the larger canals were completed and a few new ones were built, but for the most part the investors, who had waited a long time to reap any substantial dividends from their shares, were now reluctant to put up money for new ventures, and were content for the existing network to run without much attention to improvement or even to maintenance. The result was that the canals were in very poor shape to resist competition from the new railways; traffic declined, the canal companies lost money and most of them were relieved to be taken over by railway companies or to be abandoned. By the end of the nineteenth century, the canals had dwindled to a shadow of their former importance, and the decline continued remorselessly thereafter. It has been argued that, with better provision at the outset, the canals could have been built wider and deeper, and could have retained a viable role in the transport of heavy goods in conjunction with the railways, as in Germany and Flanders. This, however, is to overlook the circumstances in which British canals came into existence, and to impute to them resources which they never had. In a sense, they suffered from their originality, serving to teach other canal builders how to do the job more thoroughly. But it also seems probable that, with its insular conditions and uneven topography, Britain was better suited to a railway system than to inland waterways. Whatever the reason, the decline of the canals was rapid once the railways were established.

It is worth observing that the early railways benefited substantially from the experience of the canal engineers. The civil engineers responsible for canal construction had mastered all the techniques for cutting, tunnelling and embanking which were necessary in order to create a fairly straight and level course for a canal by the beginning of the nineteenth century, and these were appropriated by the railway builders, together with the administrative techniques of setting up statutory companies, organizing large building works and raising and directing a big migrant labour force. In several cases the same people were the engineers of both canals and railways, so the carry-over of expertise came easily. Road-building experience was also helpful to the railwaymen and, even more significantly, when passenger carriages were required for the railways, it was to the stagecoach builders that the designers turned, often employing the same builders for the task. The whole transition in transport technology which occurred in Britain in the 1820s and 1830s exemplifies neatly the ratchet view of technological innovation, with the new technology engaging directly with the mature technology and borrowing many of its basic techniques from it. The ground, in short, had been well prepared for the Railway Age which now dawned.

Although the railways were such a striking innovation of nineteenth-century technology, they were not entirely unprecedented. There had been earlier stretches of double rail, in timber or cast iron, laid down in mining operations: Agricola records some such tracks in his drawings of the mid sixteenth century. What was new was the combination of railed track with a mechanical locomotive in the shape of the steam engine. The locomotive steam engine emerged as an application of high-pressure steam. The low-pressure engine had dominated the steam-engine market until 1800, when the Boulton and Watt patents expired. For all its fine qualities, the Watt engine was too bulky to be adapted easily for locomotive purposes, even though William Murdoch succeeded in making a small steam engine drive a three-wheeled carriage. But Watt had frowned on this experiment, and it is unlikely that it could have been scaled up into a useful vehicle. It was not until Trevithick demonstrated his compact high-pressure machine at the beginning of the nineteenth century, and proceeded to apply it to a locomotive at Penydarren on a stretch of colliery tramway in south Wales, that the steam locomotive became a viable project. Even then, Trevithick encountered formidable problems, with cast-iron rails cracking under the impact of the locomotive, which caused the temporary abandonment of the system, and it was only in the second decade of the century that the locomotive began to make real progress. Again, it was in a colliery district that the main development came: in this case it was the British Tyneside coalfield where the significant developments

133

occurred. There, George Stephenson and other Tyneside inventors introduced crucial improvements, turning the steam from the cylinder into the exhaust from the furnace in order to increase the blast, and strengthening the track with stone 'sleeper' blocks at frequent intervals, so that Stephenson was able to put his engine *Locomotion Number 1* into commission on the Stockton and Darlington Railway when this opened in 1825. Four years later, the success of Stephenson and his son Robert at the Rainhill Trials of 1829 established their locomotive *Rocket* as the prototype for the Liverpool and Manchester Railway. When this opened in 1830, it was the first fully-fledged railway in the world, with passenger and freight trains operating to a regular timetabled schedule. The Railway Age had begun.[3]

Rocket, with its inclined cylinders placed outside the frame, represented a transitional stage between the early locomotives, with their cylinders placed vertically in the body of the machine, to the even more efficient locomotives which the Stephensons began almost immediately to produce in large numbers for the early railways of the world. In these machines, the cylinders at last adopted the horizontal mode which became the general practice. At first, they were placed inside the frame beneath the boiler, driving on to a cranked axle, but subsequently they were placed outside the frame, connected to the driving wheels by external cranks. In this form the steam locomotive was launched upon a century of fruitful development, with no significant departure from the system devised by George and Robert Stephenson. Engines increased massively in size and power, but the steam locomotives currently being produced in the People's Republic of China, one of the last bastions of steam power on railways, are still of the same basic type as those being built by the Stephensons in the 1830s.

The Stephensons also determined the standard gauge of railway track, adopting the 4 ft 8½ in. gauge of the Tyneside colliery track where they conducted their early experiments. As they were called upon to engineer several of the early British railways, such as the London and Birmingham Railway, they ensured that this became the standard gauge for all subsequent track. Of course, in the free-market circumstances in which railways developed in Britain, there was at first no compulsion for companies to adopt the same gauge, and several were persuaded to choose other gauges, both greater and smaller. But the only national railway to depart from the Stephenson standard was the Great Western Railway, from London to Bristol and the south-west. Under the influence of its brilliant young engineer, I. K. Brunel, this was built on a 'broad' gauge 7 ft in width. Brunel aspired to run an express passenger service rather than a glorified colliery railway, and he calculated that, in order to achieve this he required not only a very level course, which he established over most of the main line, but also a wide gauge which would compensate for the poor springing available on the early rolling stock. His objectives were thus reasonable and he was remarkably successful in fulfilling them, but as part of a national network the break of gauge between GWR track and other lines quickly came to seem a commercial disaster. The government reluctantly intervened to prevent other main-line railways from following the GWR deviation, and the pressure of competition from its neighbours and rivals eventually obliged the GWR to convert to standard gauge in 1892.[4]

The standard gauge was exported to Europe by the Stephensons, who were called upon to build some of the first track in Belgium and France, but elsewhere there was more variety: a slightly wider gauge was adopted for Ireland, and Indian railways used several gauges. To this day there are different gauges for railways in the United States and Canada, and the Australian states also have some incompatible gauges. But apart from the hazards and inconveniences created by breaks of gauge, the railways were outstandingly successful for over a hundred years in providing a transport network which virtually encompassed the inhabited land masses of the world, which displaced road and canal transport from the dominant position they had gained by 1830, and which exerted a powerful transforming affect on the societies they served. Again, Britain was the first place to experience this transformation, because it had provided the nursery for the railways, allowing most of the new techniques to be elaborated and tested between 1830 and 1850. But British engineers were not slow to take advantage of the demand for railway-building expertise in other parts of the world and to spread the benefits of the system overseas.

Main-line railway construction in Britain only began in the 1830s, but by the middle of the century the system was already virtually complete, even though an enormous amount of infilling and duplication continued right up to the end of the century. This achievement required, in the first place, an unprecedented increase in the capital available for new investment. The great bulk of this came from private investors, and it indicates both the accumulation of wealth in the British economy by the mid nineteenth century and the caution of investors, who had previously not made their resources available for speculative investment. The railway boom was thus something of a psychological watershed in attitudes towards the use of wealth, creating a much larger and more mobile capital market than had existed before. The achievement required, secondly, a new engineering dimension. The railways were built largely by civil engineers, the successors of Smeaton, Telford and others who had built the canal network. But in the boom conditions of the 1830s and 1840s, demand outran supply, and men were recruited from other professions and industries to adorn railway prospectuses and to make them attractive to parliamentary committees.

Moreover, as the permanent way was completed there was a change in the engineering requirement: unlike the canals, the railways could not be permitted to fall into neglect, and the continuing pressing need for new locomotives and rolling stock created a new breed of railway engineers who were predominantly mechanical. True, there had been mechanical engineers amongst the early members of the Institution of Civil Engineers and many continued to be members of it. But by 1847 the need for an association of railway engineers had resulted in the formation of the Institution of Mechanical Engineers, with George Stephenson as its first president. Although by no means confined to railway men, this new institution was dominated by engineers employed by the railway companies, many of them working in the great new railway workshops which all the companies were obliged to establish in order to ensure adequate maintenance and renewal of their equipment. These railway workshops became in effect large engineering factories, developing machines and techniques of mass production which became useful in other branches of engineering such as, eventually, the manufacture of automobiles and aeroplanes.

While an access of new capital and the creation of a new engineering perspective were necessary corollaries to the boom in railway construction, various social consequences stemmed from it. Some of these, such as the new intensity of urban development and the social relationships dependent upon it, are so important that we will need to return to them later. But in the context of the advent of railway transport it is worth observing the general and far-ranging nature of its impact on society. The railways provided a new and extraordinary facility for personal transport which was appreciated not only in the advanced industrial countries, but also in places like India, China and South America. This in turn created a need for railway services such as station buildings, buffets and a supply of reading matter. By the end of the nineteenth century it is probably not too much to claim that the railways had entered into the imaginative experience of more people in the world more generally and in a shorter space of time than any previous technological innovation.

8. Transport from Steam Trains to Rockets

The ratchet view of the history of technology, whereby a mature technology prepares the ground from which innovations spring and supersede it, is vividly exemplified in several of the leading events of transport history in the modern world. The way in which the experience of building canals provided a valuable basis for the subsequent development of railway engineering has already been observed. Now we have to consider an even more momentous transition in the upward curve, measured both quantitatively and qualitatively, of the capabilities of human beings for transporting themselves and their material goods: the transfer from a transport system dominated by the steam engine to one in which the immediate sources of power are the internal combustion engine and electricity. This transition began in the nineteenth century, but most of it occurred in the twentieth century and is concerned with the emergence of the automobile and the aeroplane as the dominant forms of transport. Steam power has not disappeared, and in the form of steam turbines generating electricity it remains of enormous importance. But in its direct applications to transport it has been almost entirely replaced by internal combustion and electricity. This chapter will review, first, the application of these innovations to the railway system and marine transport, and then analyse the impact of the new technologies in their more specialized forms as motor cars and aeroplanes.

The world railway network continued to grow through the

second half of the nineteenth century, and the steam locomotive was virtually the universal source of power on this system. The only exception of any significance was the introduction of electric traction on special stretches of line. The supply of current to electric trains posed a novel security hazard and retarded any general conversion to electricity, but for certain purposes, such as use on metropolitan underground railway lines, it had enormous advantages over the conventional smoke-generating steam loco-motive, and it was adopted in London on the new Central Line in the 1890s. On existing main lines, however, steam remained supreme, and was not seriously challenged until the years between the two World Wars. Then, partly owing to the emergency meas-ures of the First World War, and partly because of a decline in the profitably of railways during the inter-war years, there was a move towards rationalizing and integrating the national systems. In Britain, this took the shape of a statutory amalgamation of the 123 existing railway companies into four groups, which were in turn taken over as a state operation in 1948. The Southern Railway was encouraged to develop an extensive low-voltage, direct-current, electric service on its London suburban routes, using a 'live' third rail to carry the current, but this remained the only British departure from the domination of steam on the railways until the second half of the twentieth century.

Elsewhere, and especially in the United States, where oil fuels were cheap, some of the leading main-line services were converted to diesel traction, which emerged as a serious competitor to steam in these years. After the Second World War, when national railway enterprises in Europe were refurbishing their systems, the favourite choice of traction power was electricity, supplied at high voltage as alternating current through overhead cables. Most of the new express services such as the French Train à Grande Vitesse (TGV) and the Japanese 'bullet' train have adopted this, and in the case

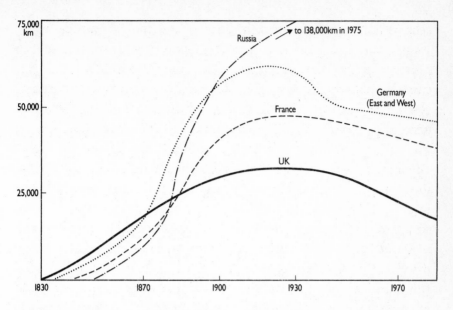

Figure 9. Length of railway line open (based on Brian R. Mitchell, *European Historical Statistics, 1750-1975*, 2nd edn, London Macmillan, 1980)

of the TGV many hundreds of miles of specially designated track have been laid to permit uninterrupted running at high speeds. British Rail have invested in sophisticated multiple-diesel units for their standard express services, but even here electric traction on the overhead wire system is being introduced as quickly as resources can be found for the installation. Although steam has thus been phased out of all Western European and North American railway services, it is generally accepted that the railway systems, with diesel or electric traction, remain essential in our increasingly crowded industrial world for high-speed inter-city transport and for the efficient carriage of many bulk commodities.

Simultaneously with the development of the railway system, marine transport also has undergone tremendous expansion and intense change over the past hundred and fifty years. It should be

141

remembered that for centuries the sea had provided the only manageable form of efficient transport over long distances, and that Western Europe had developed the sailing ship as a remarkably versatile fighting machine and freight carrier. The East India merchant ships and national naval ships of the line represented, at the beginning of the nineteenth century, a peak of technological development within the constraints of wind power and construction in wood. The application of steam to ship propulsion and the conversion to iron and steel in ship construction brought about a profound transformation of marine technology, in both its mercantile and naval forms. After several experimental efforts in Britain and France, the first viable steamship was launched on the Hudson River in 1807 by the American Robert Fulton, using an imported Boulton and Watt engine. A few years later, in 1812, the Scotsman Henry Bell began to operate a successful steamship service on the Clyde with his paddle-steamer *Comet*. All the early steamships were propelled by paddle-wheels, which may be seen as an adaptation of familiar water-wheel technology. For three decades there was little increase in the size of steamships, because a large ship posed formidable problems of fuel storage, and so they were not thought to be practical except for estuaries and inshore waters. Nevertheless, many small steamships were built for short-haul ferry and packet duties, and some of them ventured tentatively on longer voyages.

Not until 1838, however, were the problems of long-haul steamship transport overcome, when the Great Western Steamship Company began to operate a transatlantic service with the first large steamship, designed by I. K. Brunel and built in Bristol. This was still a wooden ship and paddle-propelled, but Brunel had correctly calculated that the volume of space required to carry fuel would decline in proportion to the total volume as the size of the ship increased, with the result that the SS *Great Western* arrived

142

1. Stonehenge, in the middle of Salisbury Plain, is one of the outstanding ancient monuments of Europe. But the techniques and intentions of its builders, four thousand years ago, remain largely unknown. From this vantage point, the midsummer sun rises precisely over the Hele Stone, in the centre of the picture.

2. Paddy-fields in China. Although a major world power in terms of size, China is struggling to overcome several centuries of comparative technological stagnation. However, these paddy-fields, near Guilin in south China, with a water-raising wheel in the middle distance, are well-adapted to a peasant economy.

3. Traditional farming methods still prevail in these small plots of land near Wuhan in central China; the water buffalo is pulling a very simple plough.

4. Traditional brick-making in straw-covered clamps in the open air, observed in central China.

5. Steam power. This 1788 beam engine, designed to provide rotative action and preserved in the Science Museum, London, demonstrates most of the improvements incorporated into steam-engine technology by James Watt.

6. Steam turbine. Charles Parsons's pioneering steam turbine of 1884, also preserved in the Science Museum, is shown here with its casing removed to reveal the arrangement of the rotors.

7. Atomic energy. Chapelcross Nuclear Power Station in Scotland was opened in 1959. It consists of four nuclear reactors, each with over 10,000 fuel elements, and supplies over one thousand million units of electricity each year to the National Grid.

8. Turbine hall at Chapelcross Nuclear Power Station; heat from the nuclear reactors is converted into steam to produce electricity through these turbines.

9. Old Longton. This view shows the town, at Stoke-on-Trent in the Staffordshire Potteries district, when traditional coal-fired bottle kilns were still operating.

10. New Longton. The same view showing the spectacular transformation effected by gas and electric fuels in modern kilns. A few bottle kilns have been preserved, but are no longer in use.

11. The Rocket. The original locomotive, winner of the Rainhill Trials for the Liverpool and Manchester Railway in 1829, has been preserved in the Science Museum, although it was substantially modified over a long working life.

12. The TGV. The electrically powered Train à Grande Vitesse, operating over specially designated track, has dramatically raised the performance of European railways in the last decade. This is the TGV *Atlantique* at speed.

13. SS *Great Eastern*. The third and last of I. K. Brunel's pioneering steamships, and the largest nineteenth-century vessel. It had a double hull of wrought iron and two sets of engines driving paddle-wheels and a screw. Its only commercial success was as a transoceanic electric-telegraph-cable-laying vessel.

14. *Turbinia*. Sir Charles Parsons caused a sensation in 1897 when his experimental steam-turbine-propelled launch went at twice the speed which could be achieved by reciprocating steam engine propulsion.

15. SS *Mauretania*. Cunard adopted turbine propulsion for the new breed of ocean liners which dominated transatlantic traffic in the first half of the twentieth century.

16. Dreadnoughts. Turbines were also adopted for the new type of fast, heavily armed naval vessels which were a feature of the period. This view shows units of the British Navy in manoeuvres during the 1920s.

17. Huddersfield. This aerial view of a typical British industrial town shows the close conjunction of factories and housing development. It is a Bank Holiday, so there is little smoke or traffic.

18. Hoover Dam, or Boulder Dam. Lake Mead, America's largest man-made reservoir, was formed by this spectacular concrete arch dam in 1935.

19. Combine harvester. Like every other industry, agriculture has mechanized in the last two hundred years. The internal combustion engine has been particularly important in the form of tractors and, as here, combine harvesters.

20. Early photograph. Photography was born from an understanding of light-sensitive chemicals in the first half of the nineteenth century. This view from a window of Lacock Abbey in Wiltshire is from the first paper negative prepared by Fox-Talbot.

21. 'Over London by Rail'. The French artist Gustave Doré made this realistic drawing of dingy terrace houses on a visit to London in 1871.

22. 'Rain, Steam and Speed'.
J. M. W. Turner took a rather more
romantic view of the transport
revolution in this picture of a Great
Western Railway locomotive and
train painted about 1844.

23. 'Science'. This charming piece of
Victorian iconography is one of several
decorating Holborn Viaduct in London.
It depicts an imperious female figure
holding a centrifugal governor from a
steam engine.

24. The Forth Bridge. Designed by Sir John Fowler and Sir Benjamin Baker, the Forth Bridge celebrated its centenary in 1990. It was one of the first large structures to use steel as the principal building material.

25. Sydney Bridge. The steel arch across Sydney Harbour, affectionately known as 'the coat-hanger', was built to the design of Sir Ralph Freeman and completed in 1932.

26. Golden Gate Bridge. Completed in 1937, this great suspension bridge spans the entrance to San Francisco Bay. Like all modern suspension bridges, it uses main cables of spun-steel wire.

27. Cable cars. Electric tramways gave a great impetus to suburban expansion in the twentieth century. These, in San Francisco, are unusual, in that they are driven from a continuous cable running in the groove between the lines. They are maintained now as a tourist feature.

28. Overhead tramways. The Wuppertal *Schwebebahn*, an early form of overhead electric tramway installed in 1900, still operates with modern coaches for 13 km along the course of the River Wupper in Germany.

29. Metroland. One of the most effective applications of electric traction has been in subways and underground railways. As early as the 1900s, the London Underground system represented itself as the gateway to ideal suburban homes for workers in the city.

30. Traffic congestion, 1900. Apart from the train crossing the viaduct, horse vehicles of various sorts dominate this view of Fleet Street at the beginning of the twentieth century.

31. Traffic congestion, 1960. The composition of the traffic has changed, with internal combustion engines now supreme, but otherwise the impression of slow progress through overcrowded streets remains much the same.

36. Earthrise over the Moon. The Moon was the first objective of manned space exploration, the first landing being made on 21 July 1969. Many stunning photographs were taken, such as this view from Apollo 8 in orbit round the Moon, with Earth a quarter of a million miles away.

37. Mars. Space exploration with sophisticated instruments continued throughout the next two decades. The Viking Mission put two spacecraft safely on to the surface of Mars in 1979 and took many astonishing photographs, such as this view of a thin coating of water ice on the rocks and soil.

35. Space flight. In the early days huge and very expensive, non-recoverable rockets were used, but it seems likely that space exploration will depend on the reusable American space shuttle in the future. The space shuttle Atlantis is shown here on her launch pad at Kennedy Space Center, Florida, being prepared for launch in July 1990.

33. First flight. This historic photograph caught the moment of the first controlled, powered flight on 17 December 1903. Orville Wright is lying down at the controls, brother Wilbur standing alongside.

34. Concorde. The Anglo-French Concorde is the first, and so far the only, supersonic airliner. It has operated successfully for the last twenty years. This view of the pre-production model taking off around 1970 shows its jet engines thrusting at full power.

32. The Great West Road. Efforts at traffic improvement in the shape of new arterial roads generated a great deal of ribbon development before the separation of highways from housing was accepted as the ideal arrangement and the freeways and motorways of the second half of the twentieth century were built.

38. Saturn. Voyager 2 passed by the giant world Saturn in August 1981. Its cameras revealed much hitherto unknown detail of the beautiful ring system.

39. Triton. Here, at the outer edge of the solar system, Voyager 2 took some of its best and most surprising pictures as it passed Triton, the large moon of Neptune. The photograph demonstrates the existence of an atmosphere, a polar ice-cap and active vulcanism.

40. Spaceship Earth. The vision of Earth from space has been one of the rewards of space technology. Strangely, nobody expected it to appear as beautiful and as cloud- and sea-dominated as it actually is. The spectacle has served to reinforce our sense of the fragility of the ecosphere, because this seems to be the only life-supporting environment in the solar system.

in New York fifteen days out of Bristol with plenty of coal remaining in its bunkers. This successful demonstration of the viability of larger steamships inspired a spate of shipbuilding, and Brunel was commissioned to build a sister ship to the *Great Western*. Instead, he constructed a ship which was not only substantially larger than its predecessor, but one which was also the first large iron ship and the first large ship to adopt screw propulsion. A ship incorporating so many startling innovations was bound to have problems, and the SS *Great Britain* certainly had its share, but the truly remarkable thing about this ship is that it overcame all its problems so well that it has survived a long working life to be returned eventually to the city of Bristol, where it was built, as a treasured heritage feature. It made its first Atlantic crossing in 1843, and operated very successfully from Liverpool to New York and subsequently on the route to Australia.

Brunel went on to design a third steamship, the SS *Great Eastern*. He tried to solve the problem of building a ship which would be able to carry its own fuel on a voyage round the Cape of Good Hope to India and the Far East. In order to achieve this he conceived a ship which, over 600 feet in length and displacing 22,500 tons of water, was larger than any other ship built in the nineteenth century. It was equipped with a compartmentalized double iron hull and with two sets of steam engines, one driving a single screw and the other powering a huge set of paddle-wheels. The aim of this duplication was to give the ship maximum manoeuvrability whatever the conditions of the harbour installations it would be obliged to use, and however high in the water it might be riding when carrying little cargo. The device was certainly successful as, despite its immense size, the *Great Eastern* proved itself to be capable of extremely delicate manoeuvres, which made it ideal as a cable-laying ship. From a commercial point of view, that was the only thing it did well, because it quickly became apparent that the ship was too large for the available traffic in the mid nineteenth century, and it was less efficient than the smaller ships which came into service at the same time, equipped with compound engines.[1]

The *Great Eastern* made its maiden voyage just after Brunel's premature death in 1859. By this time, new ships were at last beginning to adopt compounding as a way of increasing the efficiency of their steam engines, using the steam twice, in successive high- and low-pressure cylinders. This device had already been adopted in many large mill steam engines, but there were problems in adapting it to the more compact designs appropriate for marine applications, with the consequential increase in boiler pressures. However, once the transition had begun most non-compounded engines appeared very inefficient and wasteful and rapidly became obsolete, so the *Great Eastern* did not survive long as a cable-laying vessel. The future of steam navigation came to depend on efficiency rather than size, and compound engines were themselves superseded by triple-expansion engines in the 1870s, the steam passing through three cylinders at diminishing pressures. In theory it was possible to have any number of cylinders, and a few quadruple-expansion engines were built for marine purposes, but in practice the losses through heat exchange and friction made any further gain in efficiency very marginal and for most marine applications the triple-expansion engine became standard until the introduction of the steam turbine.

Brunel's innovations had enabled steamships to take over from sailing ships most of the traffic in goods and passengers on the lucrative North Atlantic crossing, and the advent of triple expansion made steamships competitive with sail even on the long-haul routes from the Far East such as the China tea trade, which had been effectively monopolized by the clipper ships, capable of racing home with the harvest each year. The improved steamships could not travel as fast as the clippers, but once the Suez Canal was opened in 1869 they could use a much shorter route than the sailing ships, which were liable to become becalmed in the Red Sea, and their overall punctuality and reliability gave them a substantial operating advantage. Thus the triumph of steam over sail was virtually complete by the end of the nineteenth century.[2]

However, there was little time for euphoria amongst the owners of triple-expansion reciprocating steam-engined ships. In the 1890s the steam turbine emerged as a superior form of marine propulsion, especially in larger applications. Charles Parsons had made the point that the turbine could achieve much higher speeds than any conventional reciprocating steam engine with his experimental steam yacht, *Turbinia*, which had created a sensation at the Jubilee Naval Review in 1897. Within a decade, Cunard had adopted steam turbines for their new ship the SS *Mauretania*, launched in 1906, with turbines developing 70,000 h.p. and a speed of 27 knots. The navies of the world, which at that time were in a condition of intense competition in preparation for a coming war, made a rapid transfer to steam turbines for their capital ships. And most large ships for the next generation were equipped with turbines. Triples survived in medium-sized ships until the inter-war years, and then began to be challenged by internal combustion engines. Rudolf Diesel had developed one form of his early high-compression oil-burning engine specifically for use in submarines, but it was quickly recognized that it provided an excellent power unit for any small or medium-sized ship, and it was eventually developed to challenge the steam turbine in even the largest marine applications. Virtually all large ships, indeed, are now diesel powered.

Marine transport has been transformed in other respects as well as in its power units. We have noted the introduction of iron and steel into ship construction: iron in the 1840s, and steel when it became available in bulk in the last quarter of the nineteenth century. This change had the effect of transferring the shipbuilding processes from forested estuaries and the neighbourhood of traditional ports to areas with easy access to the products of the iron and steel industry. Shipbuilding became a heavy industry, closely associated with iron furnaces, coalmines and mechanical engineering. Its central process of building iron and steel hulls and superstructure became labour-intensive and noisy, with armies of riveters creating a deafening din. This changed after the Second World War with the introduction of welding, which made shipbuilding a relatively quiet operation, but by that time other structural changes in the international market had caused a slump in the demand for new ships, so the industry fell into a prolonged decline. Part of this decline has been caused by the demise of the passenger liner with the shift to the much faster air routes, and also by competition from shipyards in developing countries, which have frequently been able to build tankers and container-carriers at a fraction of the costs of a shipbuilder in the West. But one way or another, it is unlikely that the industry has seen the end of the sweeping changes which have affected marine transport in the last century and a half.

It is almost impossible to think about the internal combustion engine without thinking of the automobile and the aeroplane, and we have already had cause to mention these when referring to the radically new pattern assumed by world transport systems in the twentieth century. These innovations, however, deserve consideration in their own right. The crucial thing about the automobile was that it revived road transport systems – which had languished in competition with the railways – and thereby promoted a massive increase in facilities for personal transport and movement of freight to suit the precise location and convenience of the customer. Road maintenance had been sustained, even when services had declined, and a large measure of responsibility had been assumed for this by national and local governments in the second half of the nineteenth century. Many European roads were recognized, after the reforms of Napoleon I, as having a strategic significance and were maintained in order to facilitate the rapid movement of armies if this should again become necessary, so that a reliable system of roads was available for experiments. It is certain that the steam carriages developed in Britain in the 1820s and 1830s by Walter Hancock, Sir Goldsworthy Gurney and others, could have been developed into viable transport systems, but the combination of mechanical problems and the opposition of the railway- and road-managing interests (the latter were alarmed by the heavy wear imposed by steam vehicles) made this difficult, and in Britain these difficulties were codified in legislation which prevented any mechanical vehicle from proceeding at more than walking speed on the roads. Thus, while steam traction engines and even steam rollers developed, no attempt was made to challenge the dominance of the railways in providing a fast transport service.

The revival of interest in mechanical means of road transport came with a surprising reversion to human power in the form of the bicycle. This had been anticipated by the 'dandy-horse' and 'velocipede', in which the gyroscopic possibility of balancing safely on two revolving wheels had been explored, and by the 1880s the 'ordinary' bicycle (or 'penny-farthing') and other types combining differently sized wheels had become popular. The 'safety' bicycle, with equal-sized, wire-spoked wheels mounted on ball bearings, the rear wheel being driven by pedals and a chain from a central axle, and with pneumatic tyres and a diamond-shaped frame of tubular steel, then developed swiftly to become the standard bicycle for the next century. It is enough for our purposes to regard the bicycle as the product of a mature engineering industry seeking to meet the rising consumer demand for a simple and convenient mode of personal transport. There can be little doubt that it helped people, and especially young people and women, to travel more widely than ever before, thereby generating a new demand for even more freedom of movement in the future.[3]

Nor can there be any doubt that the automobile benefited substantially from the precedent of the bicycle, not only in the creation of a market of potential customers, but also in terms of engineering expertise: many of the techniques devised for bicycle construction were used in car-making too, and in many cases the firms themselves switched from making bicycles to building motor cars. Whatever the precise nature of this intriguing relationship, the automobile certainly followed hard on the heels of the bicycle. Two German engineers, unknown to each other, adapted a petrol engine successfully to road vehicles; Gottlieb Daimler produced the first motor cycle and Karl Benz the first motor car, both in 1885. It does not decrease the originality of these innovations to see them as essentially inspired assemblies of pre-existing parts. The road vehicle, either horse-drawn or human-powered, already commanded a significant market. And the internal combustion engine, its engineering modelled on the steam engine, had been developing steadily since Lenoir's gas engine of 1859. The trick was to bring them together. As long as internal combustion depended on coal gas for its fuel, it remained firmly fixed to its source of supply – usually the town gasworks. The advent of oil fuels opened the way to a genuinely mobile internal combustion engine, carrying its supply in a tank on the vehicle. The petrol engine, in particular, with a carburettor (invented by Wilhelm Maybach, working with Daimler) to atomize the fuel before injection into the cylinder, held out the promise of being a versatile light engine suitable for attachment to road carriages and bicycles.

It quickly fulfilled this promise. Daimler and Benz made the necessary combinations where others had tried but failed, and their success in producing the first 'horseless carriages' was rapidly taken up and exploited by other manufacturers, especially in Germany, France and the United States. The initial response was slower in Britain, partly because of the greater commitment to the steam engine there, and partly also because of the restrictive legislation on mechanically propelled road vehicles, but amendment of the law in 1896 enabled British engineers and manufacturers to join the surge into automobiles. By the beginning of the twentieth century, nearly all the famous names of the first generation of motor-car makers were in production: Daimler and Benz in Germany; Peugeot, Panhard-Levassor and Renault in France; Olds and Ford in America; and Rover (switching from bicycles), Lanchester and Rolls-Royce in Britain. Dunlop and Michelin were producing pneumatic tyres, and Bosch, Delco and Lucas were manufacturing electrical components. The three-wheel, 'horseless carriage' appearance of the early models was abandoned, and the automobile acquired its standard form with four equal-sized wheels and the engine at the front driving through an epicyclic gearbox to a differential gear on the rear axle. Steering was by a wheel at the front seat. Efficient braking was supplied through drums or discs on the wheels, and the wheels were wire-spoked, with ball-bearing mountings and pneumatic tyres, as already developed for the bicycle. The body of the car was built on a steel chassis, and electrical fittings including the engine ignition were provided from a battery. The engine was almost invariably a four-stroke petrol engine operating on the Otto cycle, although two-stroke engines were introduced and became standard for motor-cycle use, and eventually oil-burning diesel engines were manufactured in sufficiently small sizes to make them competitive in motor cars.

This paradigmatic automobile has been the outstanding feature

of land-transport technology in the twentieth century. It has undergone countless variations in shape and size, with rear-mounted
engines, transverse engines, Wankel engines (a rotary piston engine
which enjoyed a brief popularity in the 1960s) and even gas-turbine
engines. But the basic model, as established at the beginning of
the century, remains unchanged at the end of it, and has enjoyed
remarkable stability of design. It has also enjoyed enormous popularity, which has meant that it has been big business, and therefore
eminently suitable for mass-production techniques. Henry Ford
demonstrated that the complete process of car manufacture and
assembly could be achieved within one carefully integrated operation, with its specially designed machine tools and moving
assembly-belt. His first mass-produced cars came off his production lines at Dearborn, Michigan, in 1903, and in 1908 he
began manufacturing his Model T, the 'Tin Lizzie', of which some
15 million had been built by 1927, by which time the car had
acquired a world market.

The secret of the success of Ford and of the other leading
manufacturers who followed his example in Western Europe and
elsewhere was cheapness: in order to justify mass production, the
automobile had to come within the financial reach of the people
who made it and who, with their peers in other industries, comprised the potential mass market. This had an invigorating effect
on wages and labour relationships, stimulating attention to delicate matters of personnel management within industry, and also
to devices for raising credit and to ingenious ways of advertising
the rival products. The automobile has thus become an all-
pervasive feature of twentieth-century society, transforming life in
town and countryside, as the possibilities of travelling ever further
for work or leisure have been exploited. Automobile technology,
as we have seen, matured early, and has persisted for most of the
century with exceptional stability. In the last decade of the century,

there is still no obvious rival. It is true that a serious rise in the price of oil fuels could encourage the development of alternatives to the internal combustion engine as the form of propulsion, with various electric motors or steam engines being the most obvious candidates. Both electricity and steam provided strong competition to the petrol engine at the beginning of the century, but the need to carry bulky batteries or steam-generating plant weakened their challenge. Changing circumstances, however, and especially new environmental concerns about the emission of polluting gases, could give them new opportunities. But whatever the source of power, there is no prospect of any diminution in the popularity of the automobile as a personal means of transport, with its extraordinary range and convenience.

The other great technological triumph of the internal combustion engine in the twentieth century has been the aeroplane. Here, again, there was considerable significance in the convergence of various lines of development: the experience of several generations of aeronauts who had followed the success of the Montgolfier brothers in flying in balloons since 1783 (it is of interest that both hot-air balloons and gas-filled balloons were inspired by the contemporary scientific investigation of the composition of the atmosphere and its mixture of gases); the exploration of aerodynamic principles by pioneers such as Sir George Cayley and the glider pilot Otto Lilienthal; and the development of petrol engines which were both light and reliable. But the aeroplane also involved strikingly novel elements. The first successfully controlled flight by the Wright brothers in December 1903 is one of those rare moments in the history of technology – the fruition of a completely new conception. Wilbur and Orville Wright were very receptive to past experience, both theoretical and practical, yet by carefully observing the way in which seagulls and other soaring birds achieved mastery

151

of controlled flight, and successfully seeking to imitate them, they added something distinctly different from everything that had been done before. They not only flew, they controlled their own flight.

It took some time for this achievement to become widely known, partly because of the secrecy of the Wright brothers until their patent rights could be established, but by 1908 European and American would-be aviators had seen them in flight, and had begun to imitate and improve on their designs. This was particularly the case in France, where pioneers like Henri Farman and Louis Bleriot had acquired plenty of experience of attempts at heavier-than-air flight, and where several manufacturers had developed promising engines, but British and German pioneers were also very active in the field. The result was that, within half a dozen years, experiments had been carried out with many varieties of air-frame, including monoplanes like the one in which Bleriot crossed the English Channel in 1909, and with aero-engines such as the rotary engine, which enjoyed a vogue for several years because by spinning the cylinders with the propeller it achieved a self-cooling effect. When the First World War broke out in 1914, it was not immediately clear what part the aeroplane could play other than as a means of observation and reconnaissance. However, the rapid evolution of aerial combat acted as a tremendous stimulus to the aeroplane manufacturing industry, and it also winnowed out the more robust and versatile designs from the rest of the pioneering efforts. So by 1918 a substantial aero-industry had emerged in all the belligerent countries, producing a variety of high-powered, metal-framed machines as fighters, bombers and transport aircraft.

Civilian applications could not immediately cope with the products of this industry, so there was a cut-back in production and a dispersal of expertise when the war ended. Yet the conversion of wartime bombers into passenger transporters soon demonstrated the potential for civil airlines; an apparently insatiable demand for speed in transport then gave the operators a sure market, and a new generation of airliners emerged which was specially designed for the requirements of passenger transport. American manufacturers like Douglas and Boeing took the lead in this development, and the new machines were almost invariably all-metal low-wing monoplanes with two or more engines. Biplanes had remained standard throughout the war, but stronger metal frames permitted the design of single wings cantilevered into the fuselage, with a great improvement in the aerodynamic performance. Such aircraft were also ready to replace older styles for military purposes at the beginning of the Second World War.

Although civil airlines had become well-established by 1939, especially in continental America where there was a flourishing market for high-speed transport between widely separated urban centres, the long transoceanic routes were still beyond the capabilities of land-based aircraft, even though great efforts were being put into the development of transatlantic services operated by seaplanes. Ill-fated attempts had also been made to develop the wartime dirigible airship (that is, a balloon-supported craft capable of directing its own flight) for long-distance passenger use, but both British and German experiments had ended in tragedy before the Second World War. All new experiments were interrupted by the war, and when they were resumed at the end of hostilities there was a new generation of large aeroplanes, mainly four-engined and derived from successful wartime bombers, to press into transatlantic service. These quickly became very popular, so the ocean-liner services, which had previously had this traffic to themselves, came under sharp competition and had virtually disappeared from regular operations by the end of the 1950s. By this time the aeroplane had become the indispensable means of all long-distance transport for passengers and mail, and also for many specialized commodities.

By this time, also, another development had occurred which set the seal on the reliability and efficiency of long-distance air transport. This was the adoption of the gas turbine in place of the conventional reciprocating engine. The gas turbine, or 'jet' as it became known in common parlance, is still an internal combustion engine consuming oil fuels, but it acts on radically different principles from those of the reciprocating engine. Like the steam turbine, it generates direct rotary action to drive a turbine which compresses the fuel before ignition and, in the case of the turbo-prop design, drives the propeller shaft. In the more usual form of jet engine, however, the impulse is provided by the ejection of burning fuel. The possibility of such an engine had long been recognized in theory, but the high temperatures incurred prevented the construction of a successful model until new alloys could be developed for the manufacture of the turbine blades and casing. Once these became available, the preparation of several designs went ahead in conditions of considerable secrecy because of the approach of the Second World War. Frank Whittle took out the first British patents and led the successful research team in the United Kingdom; other teams were busy in Germany and elsewhere.

Thus, by the end of the war, both Germany and Britain had pressed their first jet designs into operational service. They came too late to have much effect on the war effort, but subsequently they were adopted for most military uses because of the greatly improved performance which they gave in speed and power. The great advantage of the jet engine, from the point of view of the aeroplane, is its high ratio of power to size, which more than compensates for noise and high fuel consumption. Another specific advantage is that it enables aeroplanes to travel faster than the speed of sound, and this capability is now available to many military aircraft, although only the Anglo-French Concorde has so far adopted it for passenger transport. Nevertheless, all large passenger planes now employ jet engines, even though the emphasis has been on the development of jumbo aircraft capable of carrying several hundred passengers on long-haul flights, rather than on the attainment of exotic speeds. These aircraft, manufactured by a handful of large firms, of which the giant American Boeing company has become the acknowledged world leader, have established a reassuring reputation for reliability, though they are vulnerable to terrorist attack in a world troubled by dissident groups who are unscrupulous about the means they adopt to gain publicity.

The jet engine, which has played such an important part in the 'domestication' of air transport, making it easily available and indeed unavoidable to anybody wishing to travel long distances, was one of the significant technological developments of the Second World War. Even though the engine had been invented before the war, there is no doubt that it was the pressure of wartime competition between the leading belligerents which led to its rapid adoption as the standard power unit for aircraft. The war was also responsible for other developments in aeronautics, especially in the shape of the helicopter and the 'vengeance weapons', the V1 and V2. Like the jet engine, the helicopter had been anticipated before the war, with many experiments on rotating wing craft, including the 'autogiro', a design with a standard engine and a propeller giving forward motion which caused the wing to rotate and thus imparted lift to the aeroplane. But the true helicopter design, developed in the United States by Igor Sikorsky, came too late to make much impression on the course of the war. This had a powered rotating wing of two or more blades, and a small tail rotor at right angles to the main rotor in order to balance the torque and to assist control of the machine. Helicopters have since been widely adopted for military purposes because of their capacity for vertical take-off and landing and their general high manoeuvrability. For civil applications, however, they have been of more limited use, although they are indispensable for servicing oil rigs and for various emergency services. An alternative technique for achieving vertical take-off has been the 'jump-jet', which uses the direct downward thrust of jet engines, but the heavy fuel consumption involved has limited the application of this device to specialized combat aircraft.

The V1 'flying bomb', or 'doodlebug', and the V2 rocket bomb were both used with considerable effect against the Allies towards the end of the Second World War. The V1 was essentially a form of jet-powered pilotless plane, but the V2 was something more radically new because it involved aspects of rocket technology which had never been worked out before. There had, indeed, been elementary rocket vehicles before, such as the experimental rocket-powered automobiles and aircraft designed in Germany in the 1920s, but the capability of ascending a hundred miles and more and then descending on to a distant target with an explosive warhead was a very significant achievement.[4] Fortunately for the Allies, the German forces were not able to make full use of it before the war ended, and the techniques were then acquired both by the Americans and the Russians and formed the basis of the space programmes of the two superpowers. Thus, as an aspect of manned space flight, the rocket becomes part of modern transport development, although it will be convenient to defer our consideration of this until we look more generally at the significance of astronautics.

In concluding our review of transport technology, however, it is worth observing that there have been many other developments in the twentieth century which have yet to fulfil their potential.

Of these, the ground-effect machine, or hovercraft, pioneered since the 1950s by Sir Christopher Cockerell, is amongst the most intriguing. The hovercraft generates a cushion of air on which it floats and propels itself over land or sea. This device has enormous theoretical possibilities for ferries and river travel, but has not managed so far to displace more conventional ways of performing these operations, although a partial application in the shape of the hydrofoil – designed to skim the surface of the water – has been more popular. The possibility remains of some future application which could replace the automobile – and thereby make the task of maintaining the surface of roads and motorways obsolete. A rather similar possibility is that offered by the linear induction motor, in which the air cushion of the hovercraft is replaced by an electromagnetic effect which virtually allows a train to float along a rail, but research on this has made little progress in the last two decades.

The fact remains that transport technology has made enormous advances in the twentieth century. So much so that it would be unwise to think that the process of innovation and refinement had slowed down. Indeed, there is good reason to think that the pace of development will continue, even though it would be premature to specify what directions it might take.

9. Communications and Information Technology

While transport is concerned with the movement of goods and people, communications are the means by which people convey news, information and instructions to each other. The technology of transport, as we have seen, has undergone dramatic development since the middle of the eighteenth century. The modernization of communications started somewhat later, in the middle of the nineteenth century, even though the transformation of techniques has been every bit as sweeping and impressive as that of transport. It is true that there was an important prelude to modern changes in communication in the form of the fifteenth-century invention of printing from movable type. This was an innovation of profound significance in Western Civilization because of its popularization of books, with its consequent effects on the dissemination of information, both literary and scientific. But in more immediate aspects of communication there was little change until the nineteenth century. The organization of postal services improved in many parts of Europe, coinciding with the construction of good roads and the development of reliable road vehicles. Also, military communications were promoted during the Napoleonic Wars by the installation of semaphore telegraph systems, devised by Claude Chappe in France and imitated by British systems between London and the south-coast ports (the name 'Telegraph Hill' survives to this day at several points in the southern downlands, and relates to these semaphore signalling points). The big developments, however, depended upon the scientific understanding of the nature of electricity, and this was only achieved during the 1830s. The invention of the electric telegraph was the first practical application of this knowledge, and was followed in steady succession by the telephone, radio, television and the electronic computer. This chapter will review these stages in the transformation of modern communications, considering the nature of the interrelationships between them and their complex implications for life in the twentieth century.

We have already observed that electricity began to become available as a source of power early in the nineteenth century, first with the development of reliable chemical batteries, and later, following Michael Faraday's exploration of the relationship between electricity and magnetism, through the invention of the dynamo. The ability to produce deflections in a needle by passing an electric pulse along a wire – another use of the electromagnetic relationship – led to the invention of the electric telegraph. W. F. Cooke and C. Wheatstone entered into partnership to develop such a system in 1836 and took out their first patent in the following year. Their electric telegraph with deflecting needles was quickly adopted by the new national railway companies in Britain, which were anxious to improve methods of signalling because they related immediately to the safety of the railways, and, although cumbersome, the Cooke and Wheatstone set of dials was well suited to the mechanical signal-boxes which were then being installed. The system depended on a complicated code, and it was left to the American Samuel Morse to devise the efficient code of dots and dashes which was incorporated in the US Act of 1843 and which gave Morse powers to construct the first telegraph system there. It then performed a vital communications function in the process of opening up the American West.

Meanwhile, back in Europe, the commercial significance of the electric telegraph as a way of passing important information between widely separate markets and stock exchanges was being quickly assimilated. Cooke and Wheatstone formed the Electric Telegraph Company, which was responsible for putting up four thousand miles of telegraph lines in Britain within six years. In 1851, the first telegraph cable across the English Channel brought the business communities of London and Paris into instantaneous communication, and within a few years the links between all the capitals of Europe were complete. Nor was it only business information which benefited from this novel form of communication. News agencies like Reuters were created to exploit its possibilities, and newspapers like *The Times*, which prided itself on its readiness to adopt new technologies, used it to receive reports from its correspondents. News of events in the Crimean War in the mid 1850s was cabled immediately to London, and the public's indignation at the ill-preparedness of British troops for the conditions of the war led directly to the fall of the government. In 1870, the Prussian Chancellor von Bismarck was able to edit a telegram from the King of Prussia so effectively that it annoyed the French and precipitated the Franco-Prussian War, which he wanted.

The range of electric telegraphy was steadily extended by the laying of transoceanic cables. After several failures, the first successful transatlantic cable was laid by the SS *Great Eastern* in 1866. In many respects this enormous ship had been a serious problem to its builders and operators, but it proved itself to be ideally suited to the work of cable-laying, both because it was capable of carrying the complete length of cable required and because of its great manoeuvrability. It continued with this work across the Indian Ocean, so that in 1872 it became possible for the Lord Mayor of London to exchange telegraphic greetings with the Mayor of Adelaide, thus bringing remote cities into instantaneous contact with the capitals of the Old World. The communication was crude and basic, being reduced to a crackling code of dots and dashes, but it was remarkably effective, and the world has never seemed so large since. In Britain, the government recognized the importance of the electric telegraph system by arranging for it to be taken over by the Post Office in 1868.

The possibility of harnessing the telegraphic system to the direct transmission of speech was quickly exploited. The crucial innovation here was the discovery of a means of registering a voice on an electric current, and it was made by Alexander Graham Bell in 1876. Bell, the son of an emigrant Scotsman, was working in Boston as a speech therapist, concerned with helping deaf children to speak, and he used his knowledge of acoustics to devise a method of conveying speech by wire. However, as so often happens in the history of invention when a need becomes apparent, several other people were working towards the same objective, and one, Elisha Gray, applied for a patent only a few hours after Bell. Moreover, it required the further invention of the electromagnetic microphone by D. E. Hughes two years later to complete the modern telephone, because without this amplification device the voice currents produced in Bell's apparatus were too weak for long-distance communication. But thus fortified the invention was an instant success, and the Bell Telephone Company was set on course to become one of the largest enterprises in the world. Americans were particularly eager to develop a nationwide telephone system for business and personal communication, but it was taken up with enthusiasm in Europe also. In Britain the government once again stepped in and made the General Post Office responsible for running most of the national system. By 1958, the GPO had around 7 million telephones, amounting to about one for every seven inhabitants of Great Britain. This compared with 67 million telephones in the USA at that time, giving about one for every two and a half inhabitants.

The next leap in the development of communications was even more sensational, because it involved abandoning the inter-connecting network of wires to create 'wireless' or radio communication. Radio derived from the investigations of nineteenth-century theoretical scientists, and especially those of the Scottish physicist J. Clerk Maxwell, who had explored the relationships between light and electromagnetic phenomena in the 1870s and made predictions about the feasibility of using these to transmit messages. Maxwell's deductions were confirmed in 1885 by the experiments of the German Professor Hertz, who produced an electric current in a circuit tuned to, but separate from, a 'transmitting' circuit. Other scientists pursued these investigations, but it was left to the brilliant Italian engineer G. Marconi, working in Britain with the cooperation of the GPO, to make wireless telegraphy into a practical commercial proposition. His most impressive achievement was in December 1901, when he succeeded in sending the first wireless signal across the Atlantic. Soon afterwards, Marconi's Wireless Telegraphy Company built many wireless stations around the coast of Britain and equipped ships with radio apparatus. The disastrous sinking of the SS *Titanic* in 1912 was an early occasion when radio was used to get urgent aid to a ship in distress, and focused attention on the value of this sort of communication at sea.

Early wireless transmissions were coded, like the telegraph, and the signals were weak. The key to further progress was provided by the thermionic valve (the 'tube', in American usage), the development of which made possible the amplification and rectification of signals, so that voices and other sounds could be conveyed to the ear as broadcast by a modified telephone receiver. The thermionic valve was another product of close interaction between scientific investigation and practical experience. It derived from the incandescent-filament electric light bulb, invented by Thomas A. Edison and Joseph Swan in 1881. Intensive experimentation to discover the best possible filament led to Edison's observation of the fact that, when a filament had been burning for many hours within the vacuum of a light bulb, a blackening occurred on the inside of the bulb. Scientific analysis then correctly attributed this 'Edison effect' to a random stream of free electrons from the glowing filament (the 'cathode', or negative pole) to the inside of the bulb. It was later discovered that this weak current could be harnessed through a metal cylinder placed inside the bulb (a 'diode'); when it was modulated with the aid of another plate (the bulb was now a 'triode'), it became possible to receive signals from distant transmitters and amplify them into the voices and music in which they had been originally expressed. Credit for these inventions was the subject of bitter legal contention between J. A. Fleming and Lee de Forest, and as the prize at stake was potentially massive it is not surprising that the argument was so vigorous. But from the point of view of the history of technology it is yet another example of the tendency of inventive minds to converge on the same range of problems. In any event, it resulted in the introduction of broadcasting as a popular means of mass communication in the 1920s. Wireless sets, equipped with thermionic valves, drawing current from a battery or from the mains supply, and housed in a box pressed out of bakelite, were mass-produced to meet the demands of a rapidly expanding market. Radio quickly came to play an important part in the lives of ordinary people, imparting news and information, and providing entertainment.

The same line of scientific inquiry which had explained the Edison effect eventually led to the development of the cathode ray tube, and thus of television. To many scientists in the early decades of the twentieth century, the leap from sound reproduction by wireless to the transmission of visual images seemed inconceivable. In 1926, however, the Scottish inventor John Logie Baird demonstrated its practicality. He used a mechanical method of scanning the subject to be transmitted through a series of holes in a rotating disc; it was a cumbersome method, but it could be made to work provided the subject did not move too quickly, and it was seriously considered for public transmission services in Britain. But once Baird had demonstrated the feasibility of visual transmission, the Marconi Company in America developed a much more efficient method of electronic scanning which quickly replaced Baird's system and was generally adopted by the television services which began in Europe and America during the 1930s. They did not make much progress before the Second World War interrupted most regular services, and television made no contribution to the war. Once the war was over, however, it was soon revived to occupy a position of enormous significance in the lives and homes of most people in Western Europe and North America. Television technology has undergone continuous improvement, with the introduction of finer-definition pictures, colour transmission and the use of communications satellites to put all parts of the world within immediate visual contact of each other. It has proved itself to be a very versatile invention, with many scientific, medical, educational and industrial applications, as well as its overwhelming importance as a medium of mass communication.

The informal collaboration between science and technology which produced the thermionic valve and the cathode ray tube also produced radar, the transistor and the electronic computer. Radar is essentially a technique for bouncing radio waves off an object in order to recover the return signal and thus to give a locational 'fix' to the object. It was first developed with great success as a means of aircraft detection during the Second World War, but it quickly established itself as an indispensable navigational instrument and it has been adapted as a mapping tool, particularly for places which would otherwise be inaccessible, such as the surface of planets which, like Venus, are always covered by cloud.

The electronic computer has become one of the most distinctive and potent inventions of the twentieth century, although there were important precedents for it in the nineteenth century and even earlier. Reduced to essentials, a computer is an arithmetical machine for doing sums, and ancient devices such as the abacus, still widely used in China, were designed to do just that. In the seventeenth century, mathematical philosophers such as Napier and Pascal devised various aids to calculation, and then, in the nineteenth century, the brilliant but eccentric Charles Babbage, Professor of Mathematics at Cambridge, constructed his 'difference engine' and 'analytical engine' to perform complicated mathematical processes. Or rather, he began to construct them, because in both cases his resources were exhausted before the machine was complete, though he made sufficient progress to demonstrate the effectiveness of the mechanical operations. The trouble with these machines was that they depended upon immensely complicated chains of gearing, all of which had to be manufactured to a high degree of accuracy and assembled with precision. Babbage showed that a wide range of propositions could be translated into computer language and worked out by mechanical calculation. But it required a different and much faster technology to convert this theoretical possibility into reality.

Babbage had recognized the importance of being able to store information and programmes in the machine, and had adapted strings of punched cards as used in the Jacquard loom for this purpose in his analytical engine. A similar system of data storage was subsequently adopted by Herman Hollerith to record the American national census of 1890, providing a new type of tabu-

lating machine for use in government and business administration.[1] The storage of information was subsequently aided
by the system of mathematical logic devised by another British
mathematician, George Boole, which expressed logical statements
as a sequence of binary numbers – that is, he used only the
symbols 0 and 1. This was adopted by Claude Shannon of the
Bell Telephone Laboratories in 1938, using the two states of an
electronic component, either 'on' or 'off', to represent the two
values, thereby providing a viable basis for computer logic using
telephone relays or other electronic equipment. Scientists working
on such devices, such as Alan M. Turing of the University of
Manchester, were in great demand during the Second World War
because of their skill in deciphering complicated enemy codes, and
they were able to develop highly complex machines to help them in
the process. One of these, named Colossus, used 1,500 thermionic
valves, and when it came into operation in 1943 it was effectively
the first truly automatic digital computer. At the same time,
American scientists at Bell Laboratories developed a binary calculator using telephone relays; a team at the International Business
Machines Corporation (originally trading as the Hollerith Electric
Tabulating System, founded by Hollerith in 1890 and later known
simply as IBM) and Harold J. Aitken at Harvard were working
on the same lines, and in 1944 completed the Automatic Sequence
Controlled Calculator, designated the Mark I computer.[2]

There was a flurry of activity after the war, in both America
and Britain, as scientists explored the possibilities of developing
these machines and incorporating electronically stored programmes. All of them used thermionic valves and were necessarily
large. After 1948, however, a second generation of electronic
computers began to emerge, based on transistors rather than
valves. The point-contact transistor was first identified in that year
by John Bardeen, Walter Brattain and William Schockley at the

Bell Laboratories, as a result of work on solid-state electronic devices, and this gave a tremendous impulse to computer design. Not only could transistors perform all the functions of the thermionic valve, but they did so more robustly and reliably and, above all, on a much smaller scale. It immediately became possible to build computers of a manageable size, and the potential range of uses for them increased rapidly. Mainframes, or large-scale systems, were made available to businesses and offices in the 1950s, but these remained very expensive, and their use was controlled by the manufacturers, with the users hiring time on the machine and feeding in their own punched cards or punched paper tape.

The next big breakthrough, inaugurating the third generation of computers, came with the development of the integrated circuit in which functions hitherto requiring bulky mainframe equipment could now be performed by a small piece of semi conductor – the 'chip' – with an array of minuscule transistors printed or etched on to it. And, as well as being so much smaller, these computers were also faster. The first patent for an integrated circuit was granted to the American J. S. Kilby in 1959, and it was rapidly adopted by computer manufacturers to provide genuine desk-top computers for home and office. The computer has, in fact, become domesticated: installed in many varieties for work and pleasure, the modern PC has quickly become an important part of life, and the apparatus of expertise and industry which it represents has grown into one of the most impressive late-twentieth-century enterprises. The facility for communication with other data banks both near and far has generated the concept of 'information technology' as a description of the whole system whereby anybody may have access, through their personal computer, with mainframes containing a rich variety of information. In some advanced industrial countries, for instance, the facility exists for a wide range of services (bank transactions, airline bookings, shopping orders, etc.) to be dealt with instantly over the telecommunications network by means of a PC. There seems little reason to doubt that such services will become even more extensive and generally available in the next decade, even though abuse of the system through hacking and other sophisticated malpractices has become alarmingly common.

In pursuing the transformation of modern communications through the application of electricity, we have followed the main lines of development from the electric telegraph to the computer. But there are other aspects of the transformation which deserve mention on account of their contribution to the great facility for communication which we enjoy today. Some of these, like sound and video recording, have been promoted by knowledge of electrical science, but others, like photography, could have occurred without the application of electricity. It was that prolific inventor Thomas Alva Edison who first devised a way of recording sound. He patented his phonograph in 1877, four years before his electric light bulb, when he discovered that it was possible to record voices by using a microphone to vibrate a needle as it cut grooves in a clay-covered drum; the sound could be reproduced by running a needle through the groove and picking up its vibration on a microphone. It was a surprising discovery which nobody had anticipated, and Edison envisaged its main application as being in business. It was some time before the great potential of the gramophone, as it became more generally known, was recognized, but then it became immensely popular as a means of recording all sorts of music for leisure enjoyment. The apparatus underwent continuous improvement to the high-quality pick-up and compact disc in use today, but the principles would be immediately recognizable by Edison even though his preferred method of recording on drums has been abandoned in favour of discs.

The record-player is an acoustic rather than an electromagnetic instrument, but electricity became an invaluable adjunct in providing the driving mechanism and in improving the quality of the reproduction. Magnetic tape, on the other hand, was firmly related to the development of electrical power and electronics, as was videotape, which followed it. A method of using an electric current produced by sound to magnetize a steel tape was invented by V. Poulsen in 1889, and named by him a 'telegraphone'. For several decades, however, this made little headway in a market dominated by the gramophone, and it was not until the Second World War that the idea was taken up commercially in the modern tape recorder. Since then it has made great progress; it has many new applications in industry, science, education and communications, and the versatile tape cassette has taken over a large part of the leisure music-recording market. The technique of magnetized information recording has also been adapted to disc form, the floppy disk, and has become an ubiquitous feature of contemporary information technology.

A related technology, although it applies significantly new principles, is that derived from light as a vehicle for the communication of information. Ever since the theoretical work of Clerk Maxwell, the demonstration of the close relationship between light and electromagnetism opened up the possibility of using light as a substitute for electromagnetic signals. Research in the decades since the Second World War on the creation of 'coherent light', in which the otherwise random emanation of light waves is focused into a single band, led to the production of the laser (Light Amplification by Stimulated Emission of Radiation), with wide-ranging consequences. In addition to military applications in 'star wars' technology and to medical applications in delicate internal surgery, the laser is being introduced into many aspects of communications technology. Amongst other things, lasers are providing a remarkably quick and elegant means of transferring information from computer disks into a printout. And, in the leisure field, the CD player also depends on laser technology.

Photography stemmed from the recognition, early in the nineteenth century, that light produced a chemical reaction in certain substances, and research began to identify the most light-sensitive materials and to find a way of fixing the image left by a well-lit subject. The French physicist Niepce took the first recognizable photograph in 1826, using a bitumen-covered pewter plate and a day-long exposure. His assistant Daguerre took the process further with his 'daguerrotype', using a silvered copper plate and producing images of remarkable clarity. Each print, however, was a one-off, and could not be reproduced. Meanwhile, the Englishman Fox Talbot had been experimenting with paper coated with silver chloride. This produced a negative image from which any number of positive prints could be made. His first such photograph was a view from the window of his Wiltshire home at Lacock Abbey, and he patented this 'calotype' process in 1841.

Photographic skills were limited at first by the complexity of the processes and the cumbersome nature of the equipment – the crucial operation of developing the negative had to be performed immediately after exposure and required 'wet' treatment with several liquids. But the invention of a 'dry-plate' technique simplified the process by eliminating the need to carry a mobile dark room and developing equipment, and so went some way towards making photography a popular means of communication and recreation. There were corresponding improvements in the sensitivity of emulsions and in camera design, and the adoption of the new plastic material celluloid as the base for photographic film by the American George Eastman led to the innovation of his Kodak camera in 1888. This made photography available to anybody with modest financial resources. The celluloid, cut into thin strips and sensitized, was rolled up and placed inside the camera. The owner then made his or her exposures, and returned the camera to the makers for developing and printing, and for loading with a new film.

From the 1890s, therefore, photography has spread rapidly, assisted by the manufacture of better cameras, more sensitive films, and the development of cheap but accurate colour-reproduction techniques. Many applications have been found for it. Few weddings or family holidays are regarded as complete today without the appropriate record in a photograph album. At a more professional level, advances in photography have made an outstanding contribution to medicine and other branches of science, to various industrial techniques, and to warfare. Of more immediate relevance as a means of communication, photography has come into general use as a method of illustrating newspapers and other printed matter. In January 1890 the *Daily Graphic* began publication, the first British newspaper to be extensively illustrated. Various processes were used for reproducing photographs, with the half-tone block, using an evenly spaced pattern of dots of varying size, as the most important. The news photograph has become the normal way of giving visual impact to information.

The camera has been adopted very successfully by other technologies – especially cinematography and television. The latter has already been considered as part of the development of electrical communications, but it is worth observing that this could not have occurred without prior improvements in camera technology, and particularly the movie camera. Pioneering experiments with a movie camera had been performed by William Friese-Greene of Bristol in the 1880s, but credit for the successful invention belongs to the French scientist Marey who in 1882 made a movie camera using photographic paper and a shutter which kept out the light while the film was being moved, and which was capable of taking twelve exposures a second. Edison, active here as in other fields of inventive activity at the end of the nineteenth century, introduced 35 mm celluloid film with perforations and toothed wheels to guide its movement, and opened his Kinetoscope Parlour on Broadway in 1894. But his equipment was clumsy and only one person at a time could watch a performance, so again the credit for the first public show should go to France where the Lumière brothers opened a cinema in the basement of a Parisian café in 1895. Their achievement was soon copied in America and Britain.

The cinema enjoyed extraordinary popularity in the first half of the twentieth century, and developed rapidly from giving small-scale performances of short films which were all that was available at first to showing the great screen 'epics' which ran for several hours and entertained large audiences. For the first three decades the 'movies' were all silent, although they were sometimes accompanied by a pianist or other sound effects. Then, in the 1930s, 'talkies' were introduced; the soundtrack was recorded on the film, and was converted into sound with a photoelectric cell and thermionic valves. Soon after this, colour film began to be used in cinematography, in place of the black and white film which had previously been universal. These advances increased the hold of the cinema on the public imagination, encouraging the growth of a large film industry at Hollywood in California and elsewhere, and generating the 'star system' whereby leading actors acquired a vastly inflated popularity as box-office draws. In the critical international situation of the 1930s and 1940s, the cinema provided a valuable service of news and information through newsreels and documentary films. It was also used extensively as a means of propaganda. In a sense, the cinema became the first of the modern 'mass media', preceding even the radio.

Since the Second World War, the rapid spread of television has caused a drastic contraction in the influence of the cinema. Social habits changed quite abruptly, with television coming to perform all the services previously provided by the local suburban cinemas, many of which have followed the music-halls into limbo or have been converted into bingo halls. Only the city-centre cinemas have continued to thrive, and even these have been obliged to undergo substantial modifications in order to do so. Newsreels have been dropped and double-feature programmes introduced, and various technological gimmicks, such as wide screens, have been installed to emphasize the distinction between themselves and the small screen of television. Films continue to be made in large quantities, but they have come to depend partly on television in order to find the widest market, and the huge financial outlay involved has tended to concentrate film-making in the hands of a few big companies. Public performances remain extremely popular in India and China, however, so that, despite the decline in its influence in Western countries, cinematography remains an important medium of mass communication.

Finally, in this survey of the transformation of communications technology since the beginning of the nineteenth century, it is worth returning to where we began in order to consider changes in the oldest of modern techniques of communication – the printed word. There was actually little development in the printing process itself during the three centuries after the publication of the first book printed with movable type – Johann Gutenberg's Bible, completed at Mainz in 1456. It has been observed that Gutenberg's success derived from the fact that he was able to combine a number of pre-existing techniques: the wooden press had been developed for wine-making; the moulding of type was modelled on well-established metal-casting techniques; the ink was made from an oil-based material used for painting; and the paper, made by pulping rags, had recently become available in Western Europe. But the persistence of his innovation was remarkable and reflects the magnitude of his achievement. Europe was flooded by books of all shapes and sizes and on every conceivable subject, and through this medium an immeasurable volume of communication was conducted between persons, both individually and in a corporate activity.

The scope of the printed word increased still further in the eighteenth century with the growing popularity of newspapers and magazines, and reading habits continued to widen in the following century with the development of elementary education and democratic institutions, and the establishment of reading rooms in public libraries. This expansion required an advance in the technology of printing, which occurred in a series of stages in the nineteenth century. The first significant step towards mechanization came in 1814, when *The Times* adopted Frederick Koenig's 'pressing cylinder', a printing press which harnessed steam power to a reciprocating type-bed to turn out printed sheets at the rate of 1,100 an hour, four times the output of the hand-press. The next step was conversion to rotary action, permitting the press to run continuously rather than intermittently, and the American R. M. Hoe produced the first satisfactory rotary press in 1844. One of his machines, capable of running off 20,000 impressions an hour, was installed by *The Times* in 1856. One more step, the use of a continuous roll of paper, led to the development of the completely automatic printing press, achieved first by W. Bullock in America in the 1860s, and by the Walters rotary machine perfected by *The Times* in 1868.

While the printing press was being transformed, developments were taking place in other sections of the printing process such as typecasting and typesetting. These also were fully mechanized by 1884, with the invention of the 'Linotype' machine. It derived its name from the fact that matrices for the letters were assembled mechanically by the operation of a keyboard and each line of type was cast in a single piece from molten metal. This meant that individual pieces of type no longer had to be painstakingly assembled by hand and then redistributed when the job was finished – the used type was simply melted down.

Familiarity with keyboard controls had been greatly enhanced by the invention of the typewriter by an American printer, C. L. Sholes. The first serviceable commercial machine was manufactured by the Remington Company in 1873, and it was quickly adopted for business purposes at a time when the growing size of industrial enterprises, and the consequent need for accurate records, made it an immensely valuable office tool. Together with the tabulating machine, the cash register, and the telephone, the typewriter effected an office revolution at the end of the nineteenth and the beginning of the twentieth centuries. Perhaps the greatest social significance of this revolution was that it hastened the flow of women into secretarial work. The 7,000 women clerks in England and Wales recorded in the census of 1881 had risen to 146,000 by 1911, but even so Britain was lagging behind America in this respect. The professional opportunities opened to women through the medium of the typewriter played an important part in the movement for social equality between the sexes.

In the twentieth century, photocomposition, or filmsetting, has speeded up printing still further, while photogravure or photo-lithography have been adopted for many kinds of pictorial reproduction. New methods of rapid duplication for office and other purposes have been developed, amongst which xerography, invented by the American Chester Carlson in the 1930s, introduced an entirely new technique for printing any number of copies of a document. This is the technique of photoconductivity, a photo-electric effect in which the electrical conductivity of certain substances is increased when light falls upon them. As with other methods of fast copying, an extensive application has been found for xerography in modern business and administrative establishments. A similar success is being enjoyed by facsimile devices for scanning pages of print for transmission by the telephone network and reproduction at the point of receipt.

The production of the printed word as a means of communication has thus undergone a transformation in the last two centuries which is commensurate with the innovations in other forms of communication based upon the exploitation of electricity. It has certainly contributed in substantial measure to the enormously increased volume of information which is now made available constantly to every individual with the technology at his or her disposal. Information technology, involving the preparation of programmes for controlling computers, the storage of vast amounts of data on magnetized tape or disc, and instantaneous interchange between widely separate systems of information, has developed within three decades from a sophisticated curiosity to becoming a commonplace in industrialized societies. Of all aspects of technological revolution, this facility for instant worldwide communication must be regarded as one of the most remarkable.

10. Infrastructure – Buildings, Bridges and Services

The transformation of industrial production, transport, and communications, like the transformation of the sources of power available to technology, has been characterized by a process of cumulative development whereby virtually every innovation has emerged out of a well-established mature technology. Such a dominant technology can stimulate new departures both by its achievements and by its deficiencies. It also provides models and precedents for new developments, and encourages them by indicating possible markets. As each new technology has come to fruition, it has thrown off dependence on its predecessors and generated its own pattern of operation and market responses. The same process can be seen at work in those aspects of the technological infrastructure, such as buildings and water supply which provide essential public services. These are all part of the macrocosm of modern technology, and will be considered in this chapter.

Buildings are a primary product of any society aspiring to increase its prosperity. They are necessary to accommodate the population and to provide protection for industrial processes and machinery, and they are also essential for defensive and cultural purposes. In any archaeological examination of a society, the existence and quality of the buildings is excellent evidence for determining the level of technical development of the society. In early societies, the buildings were usually made of wood, turf, mud, and loose stone, and had very short lives. Gradually, however, techniques for working stone were acquired and buildings were constructed in well-shaped masonry: the builders of the remarkable but unknown society which created Stonehenge, for example, had mastered most of the principles of large-scale stoneworking, and the Inca civilization in the Andes of South America also achieved an impressive mastery in the use of hard stone. Stonemasons have generally avoided hard rocks like granite, preferring the more easily worked sandstones and limestones of which most of the great cathedrals and castles of Europe are constructed. But as a general rule, they were obliged to use the materials which were near at hand, because in an age of poor transport it was extremely slow and labour-intensive work to move stone over any distance.

This dependence upon local materials was alleviated by two technical developments: one was the arrival of improved methods of transport, especially by water, and the other was the spread of techniques for making artificial building materials such as brick, tile, and glass. Bricks made of roughly baked mud date from antiquity in the riverine civilizations of Mesopotamia, and they are still used in abundance in China today. The Romans devised kilning to burn their bricks more thoroughly, making them harder and more durable, and these techniques were revived in Western Civilization in the Middle Ages, so that brick and tile were used increasingly, especially in town houses. Like so many other processes, brick and tile making were organized on a thoroughly industrialized basis and mechanized in the nineteenth century. Continuously fired kilns maintained a steady flow of high-quality brick to the builders of nineteenth-century houses and factories, churches and public buildings, and as transport facilities improved with the spread of the railways, the production of brick became concentrated in areas with the best clay and access to the main markets.

The manufacture of glass also had its roots in the ancient world, but only in the sixteenth century did it become available in sufficient quantity to make the provision of domestic windows feasible. Techniques for making high-quality 'crystal' glass, as well as the simpler material used for windows and bottles, was revived in Western Europe in Venice, and spread to Flanders and then to England. The basic raw materials – sand, and wood for fuel – were widely available, but the key locational factors became the demand of growing markets and the presence of skilled craftsmen. English manufacturers evolved a distinctive cone-shaped factory, with the furnace in the middle, together with the pots of molten glass, and plenty of space around it for the glass 'blowers' to exercise their skill. These processes have become mechanized in large-scale glass production in the last two centuries, and the manufacture of modern 'float' glass, in which the sheets of glass are formed on a bath of molten tin, is now performed under carefully controlled conditions in a totally enclosed vessel.

Of all the materials which have transformed modern building techniques, none has been more important than iron. The relative abundance of iron, particularly in the form of cast iron, became a feature of Western Civilization after the introduction of the blast furnace into European technology in the fifteenth century. We have already suggested that, by analogy with the archaeological distinction between the Old and the New Stone Age, it is reasonable to regard this innovation as marking a shift from the Old Iron Age, in which iron was produced in small 'blooms' about the size of a football and was used for weapons, tools, and important implements, to a New Iron Age in which, as cast iron from blast furnaces, it became more plentiful and therefore less expensive, so that it could be used for more everyday purposes. Iron certainly became a very important building material in the eighteenth century, though rarely as an external feature of a building, in either its structure or its cladding. One building in which it appeared to dramatic effect was the Crystal Palace of 1851, with its framework of cast-iron pillars and beams modelled on the conservatory which Sir Joseph Paxton had built at Chatsworth House. There were also some churches and other public buildings assembled completely out of cast-iron sections; at least one Bristol firm maintained a thriving trade exporting prefabricated buildings to the colonies.[1]

More significant than these, however, was the application of iron to the framework of buildings to make them fireproof by avoiding the use of all inflammable materials in their construction. Such designs were introduced in the late eighteenth century and were widely adopted for textile mills and other large public buildings. Their essential features were a framework of cast-iron pillars and girders encased in brick or masonry, and brick vaulting between the girders carrying the floor above. Timber was only used – if at all – in the roof beams; even the window frames were cast in iron. Despite all these precautions, some such buildings managed to catch fire because of the processes and materials contained in them; textile raw materials and fabrics, and oily machines collecting debris and generating frictional heat, were always serious fire hazards. But in making possible the construction of large buildings in which the risks could be minimized, cast iron made an invaluable contribution to public safety, and some of the early fireproof factories have survived intact to the present.

When steel superseded cast iron as the most abundant product of the iron industry, it became available for the framework of buildings just as it did for railway track and ship construction. Being stronger than cast iron and less brittle, it could be used economically in even larger structures and thus initiated the age of the skyscraper. The first high-rise steel-frame buildings appeared in Chicago towards the end of the nineteenth century, but the style immediately became enormously popular in American cities and was then adopted – with varying degrees of enthusiasm, because some older capital cities like Paris resisted the intrusion of an alien style – throughout the Western world. Strangely enough, however, it was Paris which adopted a wrought-iron girder structure – the Eiffel Tower, in 1889 – as the dominant feature of its skyline. The outstanding image of the new townscape made possible by steel framing is the Manhattan skyline of New York City, but no modern city is without examples of this ubiquitous technique. Steel is now used increasingly in conjunction with concrete, as reinforced concrete, where steel is encased in concrete, or as prestressed concrete, where steel wires are pulled into high tension and set in concrete. The use of these materials has made possible reductions in the overall weight of structures, with corresponding increases in elegance.

Even before iron had been accepted as a major element in buildings, it had been used in the first iron bridge, constructed across the River Severn near Coalbrookdale in Shropshire by Abraham Darby III in 1779. This bridge has survived to become one of the outstanding industrial monuments in the world. As with other striking innovations, the builders worked within the parameters of known techniques: the cast-iron segments were assembled with slots and wedges like a piece of carpentry, and the shape adopted was that of a conventional semicircular masonry arch. Masonry arch bridges were indeed being transformed in Europe at this time by the adoption of more slender elliptical cross-sections, as in Thomas Telford's graceful Over Bridge across the same River Severn at Gloucester, but the innovators of Coalbrookdale accepted the orthodox shape for their iron novelty, as did most of the early iron bridge builders. It soon became apparent, however, that the iron girder made the arch redundant; when the use of a simple cast-iron girder was found to be unsafe when used in railway bridges in the middle of the nineteenth century, it was quickly replaced by composite girders built up of a web of wrought-iron members. The spread of railways gave a tremendous boost to this type of iron bridge in all parts of the world.

Meanwhile, for very large spans, another type of bridge evolved in the nineteenth century – the suspension bridge. There were long-standing precedents for this sort of structure in China and elsewhere, but as a substantial bridge slung from chains of wrought iron links the suspension bridge was an innovation of the 1820s, with Telford's bridge carrying the Holyhead road across the Menai Straits to Anglesey opening in 1826. British engineers went on to build multiple-span chain suspension bridges across the Danube at Budapest and across the Dneiper at Kiev. Then in 1855 the American engineer John Roebling used steel cable spun on site, instead of the wrought-iron chains of earlier suspension bridges, for his bridge across the Niagara gorge. He and his son Washington A. Roebling went on to build the Brooklyn Bridge across the East River in New York. This has a main span of 470 m (compared with 21 m for the Coalbrookdale Bridge and 180 m for the Menai Bridge) and was completed after fourteen years of work in 1883. Steel cable has been used in all subsequent large suspension bridges, such as the Golden Gate Bridge in San Francisco (1,280 m, completed in 1937), the Humber Bridge (1,410 m, completed in 1978) and the Japanese Akashi-Kaikyo Bridge (1,780 m, completed in 1988). It has also been adopted in the cable-stayed type of bridge in which the bridge platform is supported by cables radiating from a central tower, and which has become popular for many shorter-span bridges in recent decades.

Iron and steel have been used in other ways in bridge con-

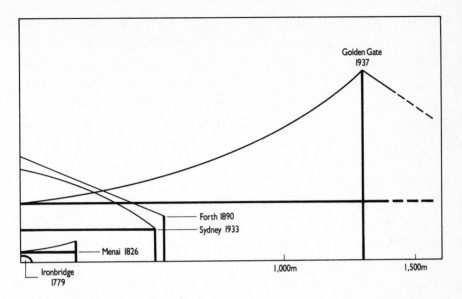

Figure 10. Large bridges of the world – comparative width of main spans (dimensions from H.S. Smith, *The World's Greatest Bridges*)

struction. Suspension bridges were not generally regarded as sufficiently rigid to perform as railway bridges, but the pressing need for larger railway bridges provided an incentive to experiment with other novel designs. The first box-girder bridge was the wrought-iron tube built in two spans by Robert Stephenson in 1849 to carry his railway over the Menai Straits, a mile away from Telford's earlier crossing point. In this design, carefully calculated and with models tested to destruction by William Fairbairn, the traffic ran through the tube rather than across the top of it, as has become the common practice in modern box-girder bridges. While this, the Britannia Bridge, was the first to rely entirely on wrought-iron, the Forth Railway Bridge of 1890 was the first large bridge to use steel. It was also the first large cantilever bridge, with the carriageway built out from three huge supporting towers. The designers, Sir John Fowler and Sir Benjamin Baker, had been made anxious by the disastrous collapse of the first Tay Bridge in 1879: this had been made up of a series of wrought-iron lattice-girders riding on cast-iron columns, and had proved to be no match for the ferocity of the gales of the estuary, one of which had carried away several sections of girder together with a complete train which was crossing at the time. The builders of the Forth Bridge thus gave enormous solidity to their novel design, and it has endured well.

As with buildings, bridge design has been greatly influenced by techniques combining steel with concrete. It has become common to prefabricate large prestressed concrete units and then to assemble them on the site. The technique was developed by the French engineer Eugène Freyssinet, and it was used for replacing many of the bridges destroyed in the course of the Second World War. Prestressed concrete has subsequently been used very extensively in the many standardized but graceful bridges on the international network of motorways. Although not attracting widespread attention as bridges, viaducts have made it possible for modern highways to pass through cities with the minimum amount of dislocation. However, their intrusion on the urban environment has been considerable, and the volume of traffic generated by them has posed other problems for overcrowded road networks.

Long experience of mining for metals and coal has provided familiarity with techniques of tunnelling and sinking shafts, and there are precedents from the Mesopotamian civilizations of tunnels made for water supply and irrigation purposes. But it is only in modern times that tunnelling has assumed importance as an adjunct to transport technology, in making it possible to carry road, canal and rail traffic through mountains and under water. The early canal tunnels were literally an extension of coalmines, as the first Bridgewater Canal ran into the colliery workings at Worsley, north of Manchester. All the excavation was done manually, but Brindley ensured that the tunnels were well lined with bricks to maintain stability. The first step towards mechanization of tunnelling came in 1825, when Marc Isambard Brunel began work on his tunnel under the Thames between Wapping and Rotherhithe. It was to be the world's first underwater tunnel, and as it was being driven through London clay most of the 'hardrock' skills of the conventional miners were not applicable. Brunel applied himself to the problems involved and was inspired by the method of the wood beetle, *Teredo navalis*, which caused havoc in the wooden ships of the navy by eating the most solid timbers. Its mouth is protected by a shield, and having reduced the wood to a pulp it lines the tunnel behind it with petrified excreta. On this model, Brunel devised a cast-iron shield built up of cellular boxes, each one large enough for a man to work in, open at the back for the removal of spoil, and with a plate at the front which could be removed when work was in progress in any of the six boxes. When all the sections had been excavated, the shield could be pressed forward by screw jacks operating on the brick lining which was built up as the shield proceeded. Formidable difficulties were encountered, and the shield was actually abandoned under water for several years while new resources were raised to complete the work. This was achieved in 1843. The tunnel was designed as a roadway, but it was converted to rail with the extension of the London Underground and is still in use.

Later tunnels under London, such as those excavated for the Underground, adopted cast-iron segments instead of bricks as lining, but the railway tunnel under the River Severn, the longest in Britain, opened in 1893, was lined with bricks throughout. Rotating cutting heads, accompanied by hydraulic equipment for advancing the shield and for placing segments of lining in position, were extensively used for tunnelling through softer rocks, and were recently used on the Channel Tunnel between France and Britain. For hard-rock tunnelling, such as that which began in earnest in 1860, when work started on the first transalpine tunnel at Mont Cenis, compressed air was introduced to drive cutting and drilling tools. This required massive hydraulic compressors to maintain the supply, but it had the advantage of providing constant air ventilation in the tunnel.

As an alternative to excavating under water, some tunnels have been built in sections which are bolted together and then placed in a trench across the river bed or estuary. Richard Trevithick experimented with such a technique for a Thames tunnel in the first decade of the nineteenth century, but was obliged to abandon it for lack of funds. It has subsequently been used successfully, as in the John F. Kennedy Tunnel under the River Scheldt at Antwerp, which was completed in 1969. For all sorts of underwater work, the caisson has become a very valuable tool, enabling men to excavate a river bed or to lay foundations for bridge piers. It consists of a metal 'bell' large enough to give one or more men sufficient room to work, together with an adequate supply of air and light. However, because it is necessary to supply compressed air at some pressure in order to prevent the caisson being flooded from below, there is a real danger of 'caisson disease', or 'bends', to the workers – a danger which was little understood in the early days of the technique.

Possibly the most important of the public services which provide an infrastructure to modern society is the water supply. This involves some very distinctive techniques, including dams, reservoirs and aqueducts. The Romans had developed an elaborate system for bringing water considerable distances by masonry aqueducts and culverts to supply the needs of their cities, and especially Rome itself. Such skills had been lost in the Middle Ages, and they were only gradually recovered as growing towns and cities struggled to keep pace with the needs of their citizens. Rivers and streams were the most obvious source of supply, once the immediately available springs and wells had been harnessed, and until the nineteenth century efforts focused on increasing the yield from this source. Then city dwellers began to look farther afield, seeking supplies from hitherto under-used rivers and impounding them to provide reservoirs. Most early dams were 'gravity' dams, which is to say that they were earth embankments or masonry walls which were able to resist the pressure of water because of their weight. The first extensive municipal water-supply undertaking in Britain was that constructed by J. F. La Trobe Bateman for Manchester between 1851 and 1877. It consisted of a series of five substantial reservoirs in the Longdendale valley to the east of the city, and it relied upon large earth embankment dams of a type which has been widely used since all over the world.

More sophisticated forms of dam such as the buttress dam, with a fairly slim masonry wall supported by a series of buttresses on its downstream face, and the arch dam, in which a thin curved wall of masonry or concrete derives its strength from the action of the arch in throwing the main load on to the adjoining cliffs, have both proved very popular. The buttress dam is particularly appropriate for controlling the flow of a large river, and has been used in the Aswan dams on the Nile. The arch dam is more suitable for use in mountainous terrain, such as the dramatic Hoover Dam across the Colorado River in the United States. It makes possible a lighter form of structure than other types of dam, and has provided an excellent use for concrete which can be mixed on the site. Not all these dams are concerned immediately with urban water supply, many being intended primarily to regulate the flow of water from what would otherwise be a spasmodic and unreliable source.[2]

In places where surface water has not been available in sufficient quantities, it has sometimes been possible to exploit underground sources, or aquifers. When the geological configuration is favourable, water can be extracted from such an aquifer by an artesian well, in which the water will be forced up the borehole by its own pressure. More normally, however, it is necessary to provide powerful pumping engines to extract water from the bores or wells, and this was one of the many ways in which the large steam engine served public needs in the nineteenth century, when many were installed to feed well-water into urban supplies. Britain had a large crop of these engines, and a surprising number of them survive although few, if any, are still in service. Pumping can now be performed much more simply by compact electrical pumps, which have few maintenance problems, although the essential function of pumping water to the urban consumer remains the same. The problem now is that, with the ever-increasing consumption of water for our modern lifestyles, the aquifers on which we have relied for the last two hundred years are being depleted. Many country rivers in the south of England have now run dry, and water is being pumped from deeper and deeper levels. The same pattern is being repeated in all parts of the Western world, making the long-term security of our water supplies a matter of some urgency.

As well as establishing the source of water supply, it is necessary to provide adequate treatment and distribution services. The coarser forms of pollution are normally removed by filtration. Reliable methods of sand-filtration were developed by British water engineers in the first half of the nineteenth century, and were fairly generally adopted thereafter. These involve passing the water through a bed of sand in the base of a tank. The sand traps the larger impurities and is regularly renewed. Chemical treatment to destroy dangerous organisms, usually by the process of chlorination, has now become obligatory for most urban water-supply services, although it only came into general practice in Britain after the Second World War. Some authorities also add fluoride to harden the teeth of children against decay, but there is continuing controversy about the medication of water supplies in developed countries.

The fact that much water is now gathered at a considerable distance from the point of consumption has made it necessary to give great attention to the means of transit and distribution. Cisterns and aqueducts, either culverted or in the open, with their attendant siphons, tunnels and pumping engines, have become part of the regular apparatus of water supply, although rarely an obtrusive element in the townscape. Water pipes were traditionally made by boring holes through elm trunks and connecting them end to end. However, the expansion of the iron industry made available new piping materials with a durability and versatility which led to their general adoption: first, cast-iron pipes, flanged and bolted together, made excellent water mains for laying under busy city streets; then wrought-iron pipes, constructed by riveting curved plates with shipbuilding techniques, were frequently adopted for aqueducts in the open air. In the twentieth century, welded steel pipe has performed the same function. For instance, the gold-mining settlements of Coolgardie and Kalgoorie in the arid deserts of Western Australia are sustained by a 645 km pipe-line along which water is pumped from the neighbourhood of Perth.[3]

Today, a good drainage system may be regarded as another aspect of water supply, particularly as most large towns in the modern world have come to rely upon water-borne methods of sewage disposal. But, in the first place, drainage involved removing excess river or sea water in order to make land habitable, and the technology of achieving this objective by means of embankments, sluices and artificial waterways has made an impressive contribution to the landscape of the Netherlands and other fenlands around the coasts of Europe. In some places around the North Sea, the gradual sinking of the land is creating a long-term problem of flood prevention, so that in exceptional weather and tidal conditions even the centre of London would be at risk. The Thames Barrage, with its ten rising-sector steel gates mounted between hooded piers across the 520 m width of Woolwich Reach, was designed to give some protection against this risk and was completed in 1984.

The cities of Europe and America coped very inadequately with the removal of organic waste before a continuous supply of water became available to provide the means of removing it. So, from the middle of the nineteenth century, there has been much investment of urban resources in the construction of systems of sewers, operating as far as possible under gravity, but with pumping engines to keep the solution in motion as required. The larger sewers were usually built of brick and adopted a pear-shaped cross-section with the narrower end at the bottom, on the principle that this would ensure a stronger flow when the level was low and thus give a self-cleansing action. Feeder sewers have generally been made with a rough but robust glazed ceramic, although cast-iron sewers have also been common, and now concrete or even plastic materials are used.

The normal practice for early water-borne sewage systems, such as the elaborate network of drains and sewers built to serve the metropolis of London by Sir Joseph Bazalgette between 1855 and 1875, was to discharge their effluent into a river some way downstream from the town being served, but this became unacceptable, especially for inland towns with other urban areas depending upon the same river lower down its course. Sewage-disposal undertakings were thus obliged to consider ways of rendering their product less noxious and, if possible, recovering usable minerals from the waste. Techniques for treating sewage involving forms of sedimentation, with the addition of chemicals to speed up bacteriological decomposition, were consequently introduced, and perform a necessary, if unsavoury, function in the infrastructure of modern city life. Because they are so much out of sight, it is frequently forgotten how important these services are, and the fact that the sewers in many British towns have now functioned with minimum maintenance for over a hundred years should give any city planner some cause for apprehension.

There are many other public services which involve significant technical skills, but as the more important of these are mentioned elsewhere in this survey they can be dealt with very briefly here. Gas supply, for example, figured prominently amongst the technological innovations of the nineteenth century which contributed substantially to the quality of urban life. Invented by William Murdoch in 1792, equipment for making coal gas was first installed at the factory of his employers, Boulton and Watt in Birmingham, ten years later. It was another ten years before the first gas supply company was formed in London, but thereafter the process spread rapidly throughout the towns and cities of Britain and was exported to Europe and America. The process involved a gasworks where gas could be distilled from coal by heating it in closed retorts; it then required some further processing to make it usable for public illumination by burning it in open jets. It also involved a cylinder for storing the gas, and miles of cast-iron and lead piping whereby it could be conveyed to its point of need. As the use of gas became increasingly popular, so the size of these gasworks and their supplementary equipment grew, to become prominent features of the urban scene. The horizontal retorts which had originally been standard equipment were abandoned in favour of vertical retorts. Being taller, these made a greater visual impact, but they also allowed continuous operation instead of the batch production of horizontal retorts, and the coke which was produced as a by-product could easily be removed from the bottom of the retorts without interrupting the process of gas-making. Other by-products, such as coal tar and ammonia, became available in abundance for neighbouring chemical industries.

The virtual monopoly in urban illumination enjoyed by the 'town gas' produced at these gasworks was challenged at the end of the nineteenth century by electricity, once Edison and his imitators had begun to provide complete systems of power generation, distribution and equipment for using incandescent-filament light bulbs. The immediate effect on the gas industry was to encourage it to diversify: it began to provide gas for heating and cooking as well as lighting, and the quality of gas lighting was immensely improved by the invention of the incandescent gas mantle, patented in 1885 by the Austrian Carl von Welsbach. But the convenience and cleanliness of electric lighting enabled it to supersede gas lighting, even though the latter remained available for many town dwellers well into the twentieth century. Meanwhile, electricity supply became established throughout the industrial countries, with large generating plant usually being driven by steam turbines with coal-fired boilers, and an extensive network of pylons carrying high-tension electric current across the countryside and thus reaching isolated places which had never been connected to the town-gas supply. After the Second World War many countries were tempted to invest in nuclear-power generation, and much has been achieved by this even though the costs and long-term dangers of the process have promoted anxious second thoughts which have yet to be satisfactorily resolved. Most electric power is still generated by steam turbines, although many of them now receive their steam from oil-fired boilers and a significant proportion are able to use water power to drive water turbines. The power from these large stations is distributed as alternating current at high voltage, instead of the direct current favoured by Thomas Edison. The high voltage makes it possible to feed the current into large distribution networks, or grids, reducing it to lower voltages by local transformers for use in homes, in industry and in the street. Modern industry and domestic life have come to rely on the instant availability of electric power, and its universality has been a formative factor in increasing the mobility of industry and population.

Some other public services will be dealt with in the course of discussing the process of urbanization. It will be enough to conclude this review of the technological infrastructure by drawing attention to the wealth of specialized technical equipment which it has generated. Earth-moving equipment such as diggers and bulldozers, building equipment such as hydraulic telescopic cranes and tower cranes, tunnelling equipment such as rotating-head cutters and conveyor belts, and a wealth of lesser tools and vehicles have done much to take the manual drudgery out of the heavy construction jobs which are essential for the maintenance of the infrastructure of modern life. The innovation and development of this equipment has been a very significant contribution to the process of technological revolution.

Part Four

THE SOCIAL CONTEXT

11. Technology and People

Technology is about machines and processes, but it is also about people. In particular, it is concerned with the immediate and long-term consequences of the relationships between machines and processes on the one hand, and people in society on the other. In this chapter we will explore these relationships, both at the level of individual participation in technological innovation and at the level of the collective involvement of people in the dominant demographic trends of our society. We will be concerned especially with the role of the individual in technological change, with the technological response to rapid population growth, with the technological style of life in modern towns and homes, and with the general consequences for individual liberty which derive from technological revolution. The emphasis throughout is on the interaction between individuals and the challenges and opportunities presented by modern technology.

The primary point of interaction between individuals and the development of technology is in relation to invention. There can be little doubt that invention is important in the history of technology, because it represents the beginning of every mechanical innovation or perception of a new process. The whole range, from the most impressive achievements to the most trivial failures, has been generated in the same way, by invention. And whatever else may be said about this profoundly important but elusive process, it is certainly human. Invention stems from the creative minds of human individuals. Most people generate no significant inventions, but some people generate many, although they have little control over the conception of their new ideas, since they are essentially random and inspirational. Thomas Alva Edison, one of the greatest of modern inventors, recognized that even to the prepared and imaginative mind, the promotion of invention meant very hard and persistent work with no guarantee of eventual success.[1] A modern study of the sources of invention concluded that a remarkably high proportion of inventions still come relatively unannounced to individuals rather than to research teams employed specifically to explore various promising avenues.[2] Of course, such research teams have come to play an essential role in maintaining the dynamic productivity of large industrial enterprises, but however successful they may be – and some of them, such as the Bell Company Laboratories, which with hard work and serendipity invented the transistor, have been outstandingly successful – much of their routine work is with innovation and development rather than invention. The production of a useful discovery depends on the unpredictable human mind.

Nevertheless, however random and unpredictable, like any aspect of human creativity, it may be, there are some generalizations that can be made about the process of invention. For one thing, any invention is conceived within an existing social matrix, which to some extent determines its qualities. This is only stating the obvious, because all human creativity must be socially determined, as individuals necessarily exist within society, and without social acceptance their inventions could have no effect. But it is worth saying, if only to avoid falling into the trap of regarding inventors and their creations as operating outside a social milieu, and interpreting the history of technology as a string of inventions unrelated to their environment. Historically, some environments have been more congenial to invention than others.

We have observed, for instance, that Chinese Civilization under the control of the mandarins became less receptive to inventions than it had been in its earlier epochs, or than Western Civilization became from the Middle Ages onwards. And in the West, we have noted that Britain was more sympathetic to invention than France or any other country in the eighteenth century, so that it was the first place to experience the acceleration in the processes of industrialization which began at that time. There is thus a wide range of relative degrees of social receptivity to invention, but the possession of some amount of readiness to experiment is an essential precondition of technological development.

These social preconditions of technological development have been rehearsed before, but they are so important as the matrix within which invention and innovation occur that they are worth recalling here. They comprise the economic factors of production – adequate capital, appropriate resources of raw materials, and a necessary minimum of technical skills amongst the labour force. But they also include the human element of creativity which produces the ideas and, most important of all, a collective readiness to accept and try out new ideas when they become available. This sensitivity to invention is a compound of many social, political and cultural factors, sustained by traditions and passed on by education and training. In short, all technological development is part of this distinctive social package, and the process of technological revolution is rooted in a long-standing readiness by Western societies to adopt the total package. The readiness has fluctuated over time, and it has been stronger in some parts of Europe and North America than others, but the overall commitment has been consistently strong and has provided the necessary social environment for the spectacular technological transformations of recent centuries.

In sketching the outlines of the social package which supports technological development, it is useful first to observe that because inventions are conceived in an existing matrix of ideas and experience, this often gives a deceptively traditional character to them, particularly at first. Thus the first automobiles were visualized as 'horseless carriages', taking over all the conventional features of existing road vehicles except their motive power. Similarly, early steamships adopted the paddle-wheel as the form of propulsion through an easy bit of lateral thinking from experience with water-wheels, though it was subsequently demonstrated that the screw was a much more efficient device. Some inventions, like the typewriter, have been so novel that they have had to struggle to establish a regular form, although once established the typewriter keyboard became a familiar feature of many later inventions.

One important aspect of the social matrix of technology is the degree of freedom allowed the inventor to express unfamiliar or unpopular ideas. Traditional societies have frequently prevented individuals from introducing devices which threaten to disrupt the harmony of existing ways of operating, and even more recently inventors have sometimes found it necessary to conduct their experiments in private in order to avoid public ridicule. Sir George Cayley was inhibited in some of his pioneering work on flying machines by such social pressures, and John Logie Baird encountered rather similar incomprehension in his early work on television. It is not hard to imagine that the lot of an inventor, however inspired, would be an unenviable one in a society such as China, where the traditional hostility to innovation is likely to be compounded by bureaucratic constraints, because in this sort of social environment the responsibility for taking decisions is avoided by pushing them to higher and increasingly remote levels in the hierarchy. By contrast, the proliferation of patent laws in Western countries, giving legal recognition and protection to the ideas of inventors, however bizarre, is a powerful indication of the willingness of these societies to allow inventors to secure proper rewards for their ingenuity.[3]

Freedom to express new thoughts and to experiment with them is thus a vital part of the social environment which promotes invention. Some measure of political liberty, a degree of freedom from the constraints of class and conformity, a tolerance towards unfamiliar and even apparently bizarre points of view are all parts of the 'social package' which we are discussing. To say this does not mean that a society which is sympathetic to invention is necessarily a liberal democracy in the modern sense. Nor does it mean that authoritarian and bureaucratic societies are incapable of adopting modern technology: the experience of recent history is powerful indication to the contrary. But it does suggest that a society which encourages, or wishes to encourage, invention – as distinct from the adoption of an existing proved technology – must be relatively open, free from bureaucratic restrictions and tending towards a liberal, democratic pattern of political organization. For although other forms of society may be willing to use the fruits of modern technology, particularly those concerned with making war, they are unlikely to have that regard for individual creativity which is necessary to enable the inventor to function and to encourage his ideas.

In considering the role of the individual in technological invention we have moved, inevitably, to reviewing individuals in society, and the fact that technology interacts with people collectively may be demonstrated in several ways. Most momentously, there is a close correlation between technology and demographic trends. This operates both through the way in which populations are increased and the way they are controlled. Both are important in the modern world, but it is in the former, as an agent of population growth, that the role of technology has been more significant, because the explosion of the world's population has rightly come

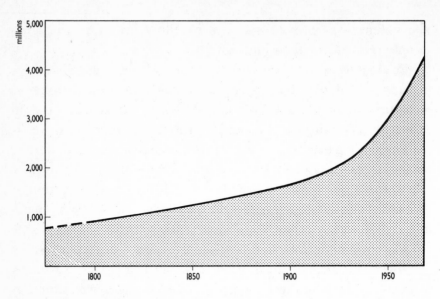

Figure 11. World population (from *Cambridge Economic History of Europe* and *Encyclopaedia Britannica*)

to be regarded as one of the most serious problems of planet Earth.

The fact of population growth is obvious enough, but the mechanisms of increase are more contentious. It has been argued that technological innovation has been a disruptive factor in human history because by encouraging population growth it has disturbed the equilibrium established between primitive societies and their environments. According to this view, societies like the Australian aborigines created an ecological niche for themselves in a hostile environment, and survived for thousands of years without significant innovation and without growth. In contrast, the societies of the Middle East which experienced the Neolithic Revolution accepted innovations in agriculture and pastoralism which multiplied the productive capacity of the groups concerned and encouraged population growth. Once started on this process, the argument continues, the pressure of more mouths to feed generated an unrelenting demand for greater productivity and consequently for more technological innovation. In the process, it consolidated the loose interpersonal relationships of simpler societies into rigidly stratified caste or class systems in order to maintain the incentives to make people work hard. Technological revolution thus assumed a major role in ensuring the stability and survival of a society which had lost its natural equilibrium.[4]

The main thrust of this argument is acceptable, even though it should be observed that the process of technological innovation was so slow in some ancient societies as to be almost undetectable in comparison with the rate of change which we have come to take for granted in the modern world. Also, the argument leaves much to be explained about the actual mechanisms whereby technology influences demographic change. In so far as primitive societies frequently controlled their population by abandoning unwanted children and the elderly who could not help themselves, while later developing societies have discarded such controls as unacceptable, it is the social mores which have changed rather than the technology. Nevertheless, several ways can be discerned in which technology has promoted longevity and thus contributed to the growth of population. Looking at modern Western society in particular, it can be claimed with some confidence, despite lengthy discussions amongst demographic historians about the relative importance of a rising birth rate and a falling death rate in causing an increase in population, that technology has had a decisive influence on the process. This has occurred through enrichment of diet, improvements in general living conditions and advances in medicine.

Dietary innovations have played an under-regarded role in the history of technology and in the well-being of human societies. It is now some years since Lynn White Jr observed that the Middle

Ages were 'full of beans', in the sense that the increased consumption of leguminous vegetables such as beans brought a
general enrichment to the medieval diet and stimulated the wealth
of political, intellectual and architectural activity which distinguished the twelfth and thirteenth centuries.[5] And this dietary
enrichment was related directly, according to White, to the new
agricultural technologies of horse power (aided by the horseshoe
and the horse collar) and the three-field system of crop rotation,
which added substantially to the total food produced and to its
variety – which is where the beans come in. However vividly
expressed, this argument is not entirely fanciful: there can be no
doubt that, over a considerable period of time and with many
setbacks, dietary standards did improve in Western Civilization,
with marginal but cumulatively beneficial effects on human vitality
and consequently on population growth.

Population growth suffered a severe check in Europe in the
fourteenth century with the Black Death and its aftermath, and
the persistence of other endemic conditions such as scurvy and of
diseases such as smallpox exercised a restraint on population
increase until the seventeenth and eighteenth centuries. It was then
that the modern wave of population growth began to gather
momentum, and once again dietary factors played an important
part in generating it. The three-field system, which had once been
a striking innovation, came to be regarded as hopelessly antiquated
by progressive farmers and landowners. It kept a third of the
productive land in fallow, and thus out of use, at any one time,
and permitted the animal stock to mix freely. In contrast, the
new intensive forms of agriculture pioneered in the Netherlands
ensured a longer rotation period with a wider range of crops, and
provided the means for careful selective breeding to improve the
stock of animals. The result was a continuing rise in agricultural
productivity and a steady flow of nutritious food to the urban

markets. Not everybody benefited equally from these improvements, but over the decades, the better diet became more generally diffused and standards of health improved. Famine, which had never been far from the consciousness of farming communities in the past, became a rarity, and by the nineteenth century had disappeared in Western Europe except for the disaster which struck Ireland in 1846. And the enrichment of diet meant that scurvy was eliminated and resistance to disease greatly enhanced. This was especially important amongst children, a greater proportion of whom survived to maturity and thus boosted the next generation of the population.

Dietary enrichments were accompanied by general improvements in living conditions. Again, the incidence of these was uneven, depending upon the resources available to different levels of the population, but average living conditions improved discernibly for Europeans and North Americans in the nineteenth century. Houses became better built as brick and tile became more widely available, with Welsh slate enjoying an enormous vogue as a roofing material in Britain, and the availability of cheap glass for windows added to the light and wholesomeness of domestic accommodation. Clothing also underwent a significant improvement in the nineteenth century, both regarding quantity and cleanliness. Good cotton underclothing became widely available through the cotton industry, while the woollen industry produced top clothes and coats of various qualities in abundance; the astonishing progress in the technology of textile manufacture paid off handsomely in bringing down the costs of clothes of all sorts, making people more ready to change them and to have them cleaned. Improved urban water supplies, meanwhile, together with the availability of good-quality cheap soap, provided a means for regular cleansing of clothes and persons, both of which had a salutary effect on health and thus on the propensity of population to increase. The improvements in living conditions were to some extent offset by the squalid conditions of many of the industrial towns of Europe, which provided a potent breeding ground for cholera and other epidemic diseases. But the devastation caused by these diseases in the middle decades of the nineteenth century provided a convincing lesson on the need for public-health measures, and these were introduced throughout most of the cities and towns of Europe and North America in the second half of the century. The epidemics of fever diseases also demonstrated that they were no respecters of social class, so that all sections of the population shared in the benefits of pure water and sanitary waste disposal once the appropriate technologies had become available.

Together with dietary factors and the improvement of living conditions, technology influenced population growth through medical provision. This operated in several ways. Medical expertise and natural drugs have existed from the earliest human societies and they were strongly represented in eighteenth-century Europe, although the cynical may doubt whether they could contribute to human longevity while remedies like bleeding patients for almost every affliction seemed to dominate medical practice. Early experiments with inoculation against smallpox, the greatest scourge of the period, were not much better, but at the end of the century Edward Jenner discovered a technique which gave genuine protection. This was vaccination, whereby the patient was given a form of cowpox from which he or she derived immunity to the much more severe smallpox. Persistence with this technique achieved the eradication of smallpox in the second half of the twentieth century – a remarkable achievement for the World Health Organization, even though the risk of recurrence can never be completely removed.

Medical techniques underwent tremendous improvements in the decades following this advance. Increasing knowledge of germs in the middle of the nineteenth century led to successful immunization against some of the worst diseases, such as scarlet fever and diphtheria; the invention of effective anaesthetics enabled surgeons to operate on hitherto inaccessible organs, although the immediate results were not encouraging, because the nature of wound sepsis through bacteriological contamination was not understood and there was a high incidence of post-surgical mortality. This problem was solved in the 1860s by Lister, building on the work of Pasteur: Lister invented aseptic and antiseptic surgery in which all instruments and materials used in the operating theatre were completely sterilized. Surgical techniques have continued to develop, with blood transfusions becoming routine and organ transplants increasingly reliable. They have been assisted by a battery of new tools and specialist skills, amongst which the application of electricity has played a very important part. Finsen demonstrated the therapeutic effects of artificially produced ultraviolet light in 1893, and Röntgen discovered the extraordinary ability of X-rays to provide information about the bone condition and internal organs of the patient in 1895. A similar line of research led the Curies to their discovery of radium in 1898, with important results for medicine as well as for other sciences. One way and another, surgical skills have made a formidable contribution to human happiness and life-expectancy, and by prolonging useful life have had an influence on the size of the population.

Meanwhile, the effectiveness of medical skills has been greatly increased by developments in the understanding of drugs. There were many time-honoured medical remedies using drugs made from plants, and some of them have been found to have a sound chemical basis, although their use was shrouded in a traditional mystique with little relevance to modern science or technology. The transformation of the old-style apothecary, dispensing such traditional drugs, into the modern scientific pharmacist occurred in the nineteenth century, being greatly helped by new techniques for separating, analysing and testing drugs. The result has been the production of a bewildering variety of new drugs for alleviating pain and almost every imaginable human malady. Aspirin (acetylsalicycic acid) was amongst the first, in 1899, and has been followed by such landmarks as the sulphonamides in the 1930s, and by penicillin and the whole range of antibiotics in the 1940s. The development of such drugs, requiring immense resources of capital and technical expertise, could only have occurred in the advanced technological countries, but their benefits have become available to the whole world and so they have become a potent factor in the growth of the world population.

Amongst the medical innovations which have had a powerful influence on modern life are some which have contributed to the restraint of population growth: the range of birth-control methods which have become available in the last few decades. Of course, birth control, both through restraint between individuals and through the use of various artificial barriers, has a much longer history. It is a remarkable fact, moreover, that as societies become more technologically advanced, they increasingly practise birth control in one form or other, so that the curve of population growth, which seemed to be following an exponential trajectory in the early years of industrialization, has flattened out in most advanced countries. The reason for this is probably the recognition that the enjoyment of the rising quality of life available in the modern world depends on limiting the number of mouths to feed in each family, but, whatever its cause, this demographic transition is one of the more hopeful signs that the population explosion can eventually be contained. For some years to come, however, the momentum of growth will continue, because the habits of birth control have not yet spread widely in the less-developed countries.

Not only has the population of the modern world grown: it has also become increasingly concentrated in towns and cities. This worldwide process of urbanization has always occurred within technological constraints, and it has been the success of technological innovation in removing many of these constraints that has enabled the staggering growth in the size of towns to take place. Towns have always been a token of civilization – an indication of permanent settlement generating trade and industry – and their existence, size and proliferation in a society can rightly be seen as a measure of the degree of the complexity and sophistication of that society. But in the early towns, the technological constraints were severe. Everybody had to live effectively within walking distance of their employment; there had to be provision for adequate supplies of fresh water; there had to be basic services for the paving of streets and the removal of waste; and fuel for light and heat had to be available. In these conditions, a normal town soon reached its optimum size, even if an imperial·capital like Rome might be enabled to grow larger by elaborate civil engineering works. There was little relaxation of these constraints until the nineteenth century, by which time the pressure of industrialization provided an urgent incentive to devise new means of allowing further urban growth. The successful development of solutions to these technological constraints may be seen in three stages.

The first stage, down to the mid nineteenth century, was dominated by the new industrialized towns which grew rapidly in Britain, northern Europe and parts of north-east America. Factory-based industries, employing a large labour force and depending upon water or steam power, tended to gravitate towards towns because these offered economies in labour supply and facilities for reaching markets and obtaining fuel. Although some of the big textile-mill complexes, like those of New Lanark in Scotland or the Merrimack mills in New England, remained in the countryside, tied to their source of water power, the natural site for a factory was a town, and the proliferation of factories gave a significant boost to town formation and growth. But most of the technological constraints on size remained in force. The workers in the factories had to live close to them because, with long hours of work, there was no time to undertake extensive journeys on foot, and no other form of transport could be contemplated by people living on subsistence wages. And the public services of water supply and waste removal, without which it was physically impossible for a town to endure, were minimal. The effectiveness of these limits was demonstrated by the cholera epidemics which ravaged the towns of Europe in the 1830s and 1840s. Even though some of the capital cities and major ports or commercial centres had managed to establish a basic water supply and to pave their streets and build some civic amenities, it had become abundantly clear that industrial towns could not expand until viable solutions had been found to these technological constraints.

Solutions began to be found in the second phase of urban growth, which ran from the mid nineteenth century to the First World War. The dominant visual feature of this phase was the coming of the railways, which were frequently imposed in a rather ruthless fashion on pre-existing patterns of urban settlement, with cuttings and brick viaducts carrying them to new railway stations in the heart of the towns. But the railways were more than just symbols of a new style of urban life. By providing relatively cheap transport facilities they materially affected the pace of urban expansion, enabling an increasing number of workers in factories and offices to live some distance from their place of employment. The railway lines generally adopted routes along the valleys leading into a town centre, and so it was along these valleys, in proximity to the railway with a series of small stations, that the

new suburban settlements sprang up. Moreover, these developments were accompanied by parallel efforts to provide fresh water
to all homes and to establish a network of drains and sewers
for the removal of water-borne organic waste. We have already
considered the importance of these public services as an aspect of
the technological infrastructure, so here it is only necessary to
observe that they acted powerfully to remove some of the most
serious constraints on urban expansion. Other public services such
as a gas-supply industry were also either installed or expanded
in this phase and, most remarkably, there was a new vigorous
expression of self-confidence in the industrial cities of the Western
world as they undertook the general paving of the streets, built
flamboyant town halls, opera houses, and other civic features, and
vied with each other to provide local transport through better
tramway services. The latter began as horse-drawn services, but
by the end of the nineteenth century most of them had adopted
some form of electric traction to provide electric tramcars.
Described lovingly by one historian as 'the gondolas of the people',
these served greatly to facilitate movement within the towns and
encouraged further expansion.[6]

The third phase of urban growth began around the First World
War and has continued ever since. It has been distinguished by a
continuing increase in the size of towns: urban areas have swelled
into each other and become conurbations; the metropolis of the
nineteenth-century capital city has become the megapolis of
modern New York or Tokyo. And the technological trigger
which has sustained this development has been power, in its dual
twentieth-century forms of electricity and the internal combustion engine. Electricity has liberated industry from reliance
on proximity to a source of fuel, because power is now available
anywhere it is required through the electric grid. Likewise, the
internal combustion engine has freed industry from dependence

on railways and canals for transport of bulky products and raw materials: they can now be readily conveyed from the factory to the customer in heavy trucks travelling along the network of roads and motorways. As a result of this removal of technological constraints, there has been a marked dispersal of industry away from the old urban centres and towards the suburbs and new towns. At the same time, the automobile has given an unprecedented mobility to ordinary men and women, who can now choose to live at some distance from their place of work and to commute in their motor cars. For those who do not have a car or who do not wish to use it for travel into the congested towns, there are fast electric-train services and motor buses, the latter having driven the electric tramcars off most British and American streets, although they are still used in some European towns.

The total effect of these twentieth-century developments has been to make towns larger and more important than they have ever been before. Most people in the developed industrial world are now city dwellers, and a high proportion of the inhabitants of the less developed parts of the world also live in cities and large towns. With the growth of the conurbations there came a rather belated recognition of the need for comprehensive town planning. There is, of course, a long and distinguished tradition of planned towns, going back to the ancient world and appearing in the eighteenth century in Edinburgh New Town and in the squares and crescents of Georgian Bath. In the nineteenth century, there were many examples of partial planning, notably of industrial estates like Saltaire in Yorkshire, but little conception of a general plan for urban life, although Paris had been comprehensively rebuilt to the instructions of Napoleon III in the 1860s. However, in the late nineteenth century pioneering prophets like Ebenezer Howard in Britain began to produce schemes for a radically new type of urban development: Howard wrote about 'Garden Cities'

and envisaged all the necessary functions of town life being performed within a new and carefully planned environment.[7] He had
many followers in the twentieth century, when persistent attempts
were made to control the rampant spread of urbanization in
ways which would create a wholesome environment, preserve the
remaining countryside and ensure the smooth movement of traffic.
Traffic engineering, in particular, has come to be a dominant
aspect of town planning, for the very practical reason that urban
congestion has increased alarmingly with the number of cars
and it has been necessary to make concessions to the internal
combustion engine in order to keep the traffic flowing freely. This
has meant the provision of urban motorways and clearways which
have frequently been less than welcome to other users of the urban
environment.

Town life has become a dominant feature of modern society,
and as such it demonstrates the intimate relationship between
technology and people. The same relationship is apparent in the
households in which people live. The organization and equipment
of the home touch everybody intimately, and they are shaped by
the available technologies: from the switch which turns on the
electricity to give light and operate a tremendous variety of domestic equipment to the furnishings and fabrics, heating and ventilating, and the video and hi-fi for entertainment. Viewed over a
period of the last hundred years, this transformation of the domestic environment is one of the most profound changes wrought
by modern technology. Yet perhaps because of its immediacy to
personal experience, it is the one which is most taken for granted.
That is a pity, because it has made a powerful contribution to the
liberalization of modern life.

Liberalization is an ambivalent concept, but it is so important
and so intimately related to the technological developments which
we have been considering in this chapter that it is worth trying to

213

place it in this perspective. The idea is difficult because it always involves freedom from something, which may change from time to time and from place to place. Thus, conditions which are acceptable and normal in one century may come to appear an outrageous imposition on the individual in the next century; social relationships which are regarded as proper between classes or sexes in one country may be condemned as illiberal or sexist in another. This makes any generalization about liberty hazardous, even though its importance in the life of modern communities can hardly be doubted. As far as the relationship between technology and people is concerned, it can be asserted with considerable certainty that an atmosphere of social liberalism is necessary to encourage inventors and to ensure them rewards from their inventions. It is also helpful in securing the dissemination and development of inventions into successful innovations. Moreover, liberalism in the form of a relatively open society, in which ideas and policies can be constructively discussed, is important in ensuring that technological innovations are properly regulated in order to prevent human suffering or environmental damage.

What we are suggesting, therefore, is that technological innovation tends to change existing relationships, often towards a freer condition than previously, although it sometimes generates new sorts of imposition which may come to be thought as undesirable as the earlier forms of illiberalism which they have displaced. While the automobile and electronic means of communication have broken down many barriers between the classes, for instance, the weapons of modern technological warfare threaten people with a greater tyranny than any they have ever experienced before. And while the typewriter, the telephone and the bicycle – as well as the labour-saving devices in the home which we have just mentioned – have contributed significantly to the emancipation of women from the thraldom of the domestic round, many observers of modern societies remain conscious of deep inequalities between the sexes, some of which have been reinforced by the traditional educational bias which gives boys a better understanding than girls of technology.

Whatever the difficulties of generalizing, it is probably true to say that most people in the industrialized parts of the world, when they make comparisons between themselves and their parents or between themselves and their counterparts in less-developed countries, feel that they are freer. And despite all its ambivalences, much of this freedom is derived from technological competence – the ability to move, to communicate, to produce and to entertain oneself. Whether or not people are equipped to take advantage of these facilities in order to become more enlightened or morally superior raises issues of profound importance which go far beyond the scope of the present discussion. Here it is enough to recognize that there is an intimate and critical relationship between technology and the freedom or otherwise with which people are able to take advantage of it. Such freedom needs to be codified in human and institutional terms between the individual and the state, so it is to this subject that we turn in the next chapter.

12. Technology and the State

When men and women are described as social animals it is meant
that they exist naturally in societies and that, without such
relationships, life is, in Hobbes's famous words, 'nasty, brutish
and short'.[1] Virtually the whole of human life is thus conducted
within a network of social relationships. These associations fall
into two categories: the obligatory and the voluntary. Most social
relationships are voluntary, in that we can choose, in theory at
least, whether or not we will belong to churches, trade unions,
conservation societies and tennis clubs. Obligatory societies are
fewer in number, but they are crucially important. There are only
two of them: the family, into which we are born, and which gives
us our ethnic and cultural orientation without our having any
choice in the matter; and the state, which is responsible for maintaining law and order within a given territorial community and
its cohesion in the face of aggression from outside. Again, we have
no choice about the state into which we are born and, such is the
way in which human communities polarize into the self-defensive
groups which are states, even if we opt out of one state it can only
be in order to move into another. There is no meaningful existence
for a stateless person. So the state is a most important form of
society, and one with enormous potential for the use and abuse
of technology.

In the first place, states are subject to the same sort of technological constraints on size which have influenced the growth of

towns. The effective power of a state to maintain law and order and the necessary means of defence against possible aggression is limited by its technological competence to raise and equip a police force and an army, navy and air force; to transport these forces easily to wherever they are needed; and to maintain communications with these forces while they are operating. Even with its superior road system and political organization, the Roman Empire pressed these constraints to the limit, and its legions frequently found themselves fighting beyond the scope of the imperial administration, so that they began to function as states within states and thus to pose a challenge to the cohesion of the empire which ultimately it was unable to resist. With the collapse of Roman authority in Western Europe, the region reverted to a simpler form of society as the invading Teutons introduced their family-based tribal states, which represented the original and basic form of state organization.

The region eventually re-established cohesion when fresh influxes of potential invaders were kept out by the increasing prosperity and strength of the settled communities, although the fact that these communities never coalesced into a single state has had a very significant influence on the development of Western Civilization. It has meant that, unlike the great monolithic state-civilizations of Egypt and Rome, the West has always presented a bewildering profusion of different countries, which have for much of the time been in strident competition with each other. This very competition, however, has provided the possibility of alternative styles of government and has enabled traditions of dissent, tolerance and liberalism to develop. This diversity has also stimulated rivalry in technology.

We have already had cause to observe the importance of liberalization in the processes of technological development, and we should now note that the capacity for such liberalization was an integral part of the relationship between many conflicting states in Western Civilization. That is not to say it was intended. Between Papacy and Empire, Catholic and Protestant, and between one monarch and a rival claimant to his throne there was little room to indulge anything resembling what the modern world would recognize as democratic liberalism. But the preoccupation with these conflicts between states in the West, as well as many others of a more personal or dynastic nature, allowed a tradition of freer and more dissenting opinion to emerge, and even encouraged it when, through access to mercantile prosperity or technological competence, it was able to promote the interests of one side or the other in these conflicts. Thus the distinctive package essential to technological revolution – liberalism, mercantile enterprise, scientific inquiry and technical skills – began to take shape, and the states of Europe had increasingly to take it into account.

One way in which states sought to encourage technologies which would give them advantages over their rivals was by granting monopolies to inventors, ensuring them of financial rewards for their ideas. These were first seen primarily as a way of promoting trade and industry by protecting specific enterprises in commercial expansion, such as overseas trade with the Indies or bringing foreign craftsmen to mine English copper. The concept was conveniently adapted to offer inventors a patent which gave them the opportunity to earn a financial reward for their ideas, by allowing them the sole use of the invention for a stipulated period. The system in Britain was complicated and expensive in the eighteenth century, and many inventors could not afford to take out a patent or chose to place their reliance on secrecy, but well-established industrialists such as Boulton and Watt made good use of it to protect their steam-engine patents for a quarter of a century. It was subsequently simplified and made cheaper, so that more and more inventors took advantage of patent protection.

Most other industrialized countries in Europe and North America developed similar systems in the nineteenth century, and patents came to acquire international recognition, although some countries were able to exploit gaps in the legislation. The most important thing about the patent system from our point of view, however, is not so much its extent or efficiency as the fact that its existence represents a recognition of individual merit, and as such it is a valuable part of the technological package.[2]

By the eighteenth century, some of the traditional constraints on technological development had begun to ease. In particular, mastery of the sailing ship and of cannon had enabled several European states to establish substantial overseas empires. It was ironic that, while inland transport remained exceedingly primitive, the sailing ships of Portugal, Spain, France, the Netherlands, England and Sweden were able to range over the oceans of the world. They could set up trading stations and colonies on the coasts of India and the New World, but it was more difficult for them to penetrate the interior of the continents behind these bases. Admittedly, the Spanish managed to do so in Central America, thanks to some exceptional bravery and ruthlessness on the part of the Conquistadores and to their luck in finding the indigenous populations handicapped by divisions and superstition. But this remained exceptional until the British enjoyed somewhat similar advantages in their conquest of India at the end of the eighteenth century. Thereafter, the persistent build-up of European settlers and fire-power made possible the progressive annexation of most of the non-tropical parts of the world. In North America, the settlers took on the job themselves, having dismissed the metropolitan authority of Britain. As far as the tropical parts of the world were concerned, imperial annexation had to await further technological tools, especially the small reliable steamship, able to negotiate rivers through equatorial forests, and anti-malarial

219

drugs such as quinine, which enabled white men to survive where previously they had died. These, together with the constant upgrading of European weaponry and the growth of the transcontinental railway networks, opened up the interiors of all the continents to European penetration.[3]

The acquisition of overseas empires by the European states had begun in the sixteenth century, at a time when the states concerned were still essentially products of medieval dynastic arrangements, although the Protestant Reformation soon added a dimension of religious antagonism to the situation. They were, it is true, beginning to think of themselves as 'nations', although it took some time for the fully-fledged modern conception of the nation-state to mature. However, the French Revolution of 1789 delivered a shattering blow to the principle of dynastic legitimacy and introduced an even more powerful principle into European politics – the voice of the people, at first the expression of an inchoate mob, but gradually gaining in coherence and self-expression with the popular mass-movements of the nineteenth century, which were often directed towards nationalist objectives. The result was that the states of Europe became nation states and adjusted their objectives accordingly. It was not so much that the states became any larger, although some of them did, nor that the condition of rivalry between them became any less – in some case it became even sharper. What was new was rather that the European states adopted nationalistic attitudes in relation to languages, ethnic origins and cultural traditions. State rivalry continued to be a stimulus to technological development, but it now came to serve nationalistic aspirations.

We will return to the distinctive contribution of technology to war between conflicting nation states, but here it is important to observe that technological developments frequently served nationalism in other capacities. The effective removal of all technological constraints on the expansion of the state, through modern means of transport, communications and generating power, has helped to promote nationalism and imperial pretensions. It has also put at the disposal of governments formidable powers of information control and propaganda which help to produce a quiescent public opinion. It is true that we have passed 1984 without fulfilling George Orwell's nightmare of a grim measure of total state control over the population through 'newspeak', 'telescreens' and other devices. But in some respects the powers available to nation-states to browbeat or hoodwink their citizens are even more alarming than those Orwell envisaged, and vigilance against the growth of such nationalist intimidation remains essential in a healthy society.

The concept of social health begs many questions unless it is stated clearly. Here it is intended to convey the notion of a social consensus based on a high degree of participation in the processes of decision making, both in the state and in more immediate levels of social action. In another word, we are talking about democracy. Whatever this much-abused word may mean to different people, 'government by the people' should certainly involve such active participation, because without it social control comes to rest in the hands of those with wealth or expertise, and 'people power' becomes attenuated, if it exists at all. So it is taken as axiomatic here that a condition of social health is a democratic form of government sustained by a high level of personal participation. It is also taken for granted that such a condition implies a well-articulated rule of law, in which the rules of the state are determined by a generally acknowledged process of law-making which includes provision for changing the law and to which all members of the society are subject.

If it can be agreed that this is a fair statement of the sort of government implied by the term 'democracy', and that such conditions are, on the whole, more desirable than any other social arrangement and are thus conducive to maximum social health, it is easy to see that there are important consequences for technology. For one thing, the resources of modern technology put enormous powers at the service of the democratic state to communicate information and to encourage participation. But the same resources can be acquired by other forms of government and used, as we have already noted in the instance of nationalistic propaganda, for less desirable purposes. The examples of twentieth-century dictatorships, of which there have been a large number of a particularly unsavoury character, have demonstrated just such abuse of the powers of technology, and it is these which give substance to the Orwellian nightmare. It is pointless to deny that technology can be used very effectively by states which are not democratically inclined, and it is important to take appropriate political precautions against such abuse. But in the last resort such abuse is alien to the concept of technological revolution advanced in this book, because it is of the essence of this process that it is part of a package and that the political part of that package is a condition of liberalization which tends towards democracy. Although there is thus a very real danger of technology being subverted for illiberal purposes, the success of such subversion must ultimately result not merely in the destruction of democracy but also in the destruction of technological revolution as we understand it. Such consequences imply the collapse of the fabric of modern society, and would be extremely sombre for the world community.

While not minimizing these dangers, it is reassuring that modern liberal democratic states can take useful precautions against them, all of which involve a recognition of the role of technology. These precautions apply in the fields of education, science, culture and war, and it is worth considering each of these in turn. The technological role in education has not always been appreciated, but it has become increasingly important in the last two centuries. Up until the eighteenth century, education had been perceived as an élite discipline designed to cultivate the minds of the ruling classes by inculcating the classics and, to some extent, mathematical theory. The acquisition of skills, either for employment or for personal satisfaction, was regarded as irrelevant to educational theory and practice, so no formal provision was made for it. It was expected that technical skills, in particular, would be acquired in the course of practice: in the case of the more ancient and traditional skills like those of stonemasons, carpenters, coopers, blacksmiths and metalworkers, this involved a degree of social formalization as apprenticeship, in which a pupil was indentured to a master for a period of years, in the course of which he would be expected to learn the craft. The parents of the pupil paid for this privilege, but in return the pupil received both training and entry into the society of the craftsmen.

It is important to recognize that this system of training by apprenticeship worked well for many centuries, during which period it represented the only form of technological education available. In view of this long tradition, it is not surprising that modern attempts to supplant it by a more institutional form of education have been resisted, nor that apprenticeship has been subjected to unnecessary derogation in the process. The fact was, however, that by the nineteenth century – if not earlier, as some prominent crafts such as architects and surgeons had already taken steps to improve and regulate the instruction of entrants into their skills – apprenticeship had become inadequate to deal with the new demand for technical instruction. These were increasing as a result of industrialization, and they were changing in quality on account of the introduction of technological novelties and scientific discoveries. The advent of the railways, for instance, created a need for more trained engineers than could be provided by the routine of pupillage in the offices of existing engineers, and the new fields of electricity and organic chemistry, areas of expertise with great technological significance, were subjects which could not be taught by apprenticeship because they were beyond the expertise of active practitioners.

The solution to these and related problems was the development of a system of technological education in schools, colleges and universities, which at first coexisted uncomfortably with the older traditions of educational practice, but which was eventually assimilated into a reasonably viable amalgam. The need was demonstrable, and when existing educational institutions could not be adapted to provide it, new bodies were devised. The French state led the way after the Revolution of 1789, setting up the École Polytechnique in 1794. Prussia followed, with prestigious colleges of technological instruction, and in the United States 'land-grant' colleges adopted technical subjects as they were established in the new states. Sweden also pioneered technological education at the university founded in 1829, which took the name of the Scottish merchant who endowed it, William Chalmers. Only Britain clung to the traditional methods of instruction, although it should be remembered that the very success of British industrialization in the early nineteenth century provided a powerful argument against change, and the mid-century predilection for *laisser-faire* in education as in other departments of public policy removed any possibility of state initiative. Even Britain, however, was obliged to recognize the need, made all the greater by the palpable successes of its rivals in the second half of the century, and began to establish new courses and new institutions in which technical instruction figured prominently. By the end of the century, therefore, virtually every state of the Western world had accepted the case for technological education and established appropriate institutions, even when this involved changing the perceptions of the role of the state and undertaking substantial educational commitments which earlier administrations would not have been able to perform. Technical education has thus become an integral part of state educational provision.

One aspect of the increasing formalization of technological education has been the rise of professionalization. At a time before any state provision for education, groups of men skilled in the new technologies began to form societies to represent their interests, especially by controlling recruitment into their ranks, and to compensate for the lack of formal educational facilities by providing for the self-instruction of their members through discussion and conviviality. British engineers provide an excellent example of this trend. Hardly existing as a recognizable group before the middle of the eighteenth century, a few of them began to call themselves 'civil engineers' in conscious distinction from the military engineers who were then better known. Led by John Smeaton, they formed a Society of Civil Engineers in 1771 as a sort of gentlemen's club with some professional functions grafted on to it. Two generations later, in 1818, the number of people in the embryonic profession had grown sufficiently to sustain a fully-fledged professional organization in the shape of the Institution of Civil Engineers. Under the benign presidency of Thomas Telford this achieved its Royal Charter in 1828 and went on to become the prototype for many similar engineering institutions in Britain, the British Empire and the United States. These, and similar bodies for architects, surveyors and other new professional groups, were not created by the state, but the nineteenth-century state tended to look on them with approval, as representing the extension of the self-help and social discipline which were regarded as desirable. Professionalization has thus been generally welcomed as a by-product of modern technology, and has often established an intimate relationship with the education of the specialist group concerned, making a substantial contribution to the emergence of an élite managerial class.

The growth of technological activity in the modern world has closely paralleled the development of science, but it is necessary to recognize these as two distinct though related phenomena, as was observed at the outset of this study. Science is concerned with knowledge and with understanding the nature of human beings in their environment, whereas technology is concerned with making and doing things, that is, with the way in which human beings change their environment. While technology is as old as humankind, scientific experience has depended upon the skills of literacy and numeracy, and so is relatively modern, being no older than the early civilizations. They tended to develop separately because of divergent social qualifications: science was recognized as the business of the literate classes, be they the priests or mandarins or the aristocracy – although it received little specific recognition in the traditional educational syllabuses – while technology was associated with the craftsmen, the labourers and the slaves. Only in the slightly more open conditions of medieval society of Western Europe did a *rapprochement* begin to occur between scientific inquiry and practical skills, resulting in several important advances in time-keeping and navigation. The foundation of the Royal Society in 1662, and of similar national scientific institutions in other countries at about the same time, was strongly encouraged by the usefulness of scientific discovery: in Bacon's famous expression, it was intended to increase man's 'dominion over nature', and inventions such as the steam engine could not have occurred without the development of a fertile relationship between science and technology.[4]

In the last two centuries, however, this relationship has become even more closely symbiotic, with a growing mutual inter-

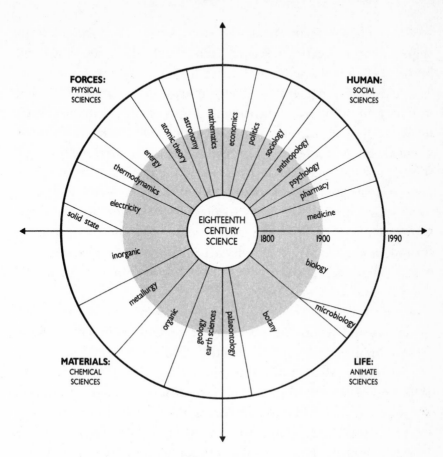

Figure 12. Science – the knowledge explosion since 1800

dependence between science and technology. Thus, the enormous success of the steam engine stimulated speculative men to explore the nature of the processes within a heat engine and to develop the theories of thermodynamics which have made such an impressive contribution to modern physics. This advance in understanding in turn promoted engineers like Rudolf Diesel to develop the internal combustion engine with his theoretically superior high-compression engine. Similar developments have occurred in other branches of science and technology, most notably in the relationships between electrical science and electrical engineering, and between chemical science and the technologies of chemical engineering and pharmaceutics. The implications of this increasing intimacy in terms of educational provision have already been noted; it is, after all, the increasing dependence of technology on scientific theory and understanding that has increased the need for formal institutionalized instruction in these fields. Yet the two activities retain a distinct identity. In particular, the development of scientific knowledge tends to proceed in jumps or 'paradigm shifts', while technological experience is more cumulative, relying on the ratchet mechanism which has been expounded in these pages. There are good practical reasons, therefore, for treating them separately.

By cultivating technological education and the interdependence of science and technology, modern states can do much to create and maintain conditions of social mobility whereby individuals can become conscious of opportunities for self-improvement and aware of a substantial personal stake in society, thus reinforcing the foundations of liberal democracy. Much the same applies in relation to the state and more general cultural activities such as literature and the arts, religion and public opinion. It is not that a high level of state intervention is either needed or desirable in these areas, but rather that a state which is disposed to tolerant passivity here will do more to promote the potential advantages of technological development than one which seeks by aggressive promotion or intolerance to direct the course of cultural transformation. In other words, any *dirigiste* attempts by the state to impose a rigid censorship on the arts, to favour any one religious faith at the expense of others or to mould public opinion by high-powered advertising and propaganda, are undesirable not only because they tend to undermine the basis of liberal democracy but also because they sap the sources of artistic and technological creativity. It needs to be remembered that the precious springs of technological innovation derive from the same reserves of creativity which stimulate other aspects of cultural activity, and to restrict the one is to restrict the other.

The close relationship between technology and general cultural creativity is apparent in all sorts of ways. It is true that technology has not figured prominently in the traditional literature of Europe and North America, but there is a growing literary familiarity with the apparatus of high technology in science fiction, and although much of this has still to be regarded as pulp literature, some of it is generating a stream of good writing which deserves to be taken seriously both as an art form and as a medium for significant speculation about the essence of technological society. In the visual arts, technology has been making an impact for a long time, in the architecture of railway stations and concrete bridges, for instance, and in such artistic movements as Impressionism, Art Nouveau, Cubism and Art Deco. Music has been less obviously influenced, although modern musicians are grasping the opportunities provided by new techniques and instruments to widen their range, and to synthesize and record their creations. The phenomenon of pop music, providing powerful yet balanced amplification to massive audiences, has been an outstanding expression of this relationship, and one, incidentally, posing important questions of social control. While it is in the public interest to exercise a measure of discipline over the timing and location of pop festivals, it would be a serious error to use such minimal control as a means of directing the form of such events or of making it impossible to hold them. The relationship is often a sensitive one, but here as elsewhere the ideal is one of minimum state intervention and maximum individual self-expression.

The same applies in relation to technology and religion. At first glance, there is no obvious connection between them. Religion has traditionally been the subject of intense interest and activity by the state, which has frequently directed citizens in their observances while forbidding any alternative expressions. Such interventionism has now been virtually abandoned by all modern Western states, even though quite a number maintain a semblance of a national church. It has been found by bitter experience that toleration is the only relationship that can be happily sustained between religious faiths which are superficially incompatible, and the lesson is one which has had far-reaching implications for cultural creativity, liberal democracy and technological society. In so far as religion is concerned with the goals or objectives of human life, moreover, there is a close correspondence between it and the material conditions which are made available through technology. Technology can bring both material wealth and physical destruction, and this poses problems which all religions in the modern world must consider very seriously. Once again, an environment of social tolerance is necessary for this.

Another, even more general area of cultural transformation is that of public opinion. This is not a precise concept, like membership of a particular church, but it is of great importance in any democratic society. It represents a consensus of feeling amongst a predominant part of a society, and implies a measure of participation by all members of that society in the processes of decision-taking. It is not appropriate, for instance, to speak of public opinion in a dictatorial state, in which the bulk of the members are never consulted on any matter of public policy, nor is it appropriate in a society which does not possess any means of communication whereby public opinion can be formulated and expressed. By making available the means of rapid and easy communication, through the printed word and through telecommunications and television, modern technology has made an outstanding contribution to the development of a genuine public opinion in Western nations. In this way it has played a part in the growth of democratic institutions and thus, in a fundamental sense, in promoting a healthy society.

We have, finally, to consider the relationship between technology and the state in the matter of warfare. This is an area of the greatest significance to the integrity and even the survival of the state, and it is one in which technology has made a long-standing and increasing contribution. The possession of weapons superior in quantity and quality to those of rivals has been an objective of states since antiquity. We have already noted how the guns and sailing ships of the European states enabled them to establish vast maritime empires from the sixteenth century, and how the subsequent improvements in technology made possible the expansion of these European empires in the nineteenth century. Amongst the nation states of modern Europe, however, there has been remarkably little technological imbalance, as every state has struggled to copy, keep up with and get ahead of the weapons of its rivals. The result has been a condition of permanent armaments race, even though the tempo of this constant rivalry has been muted from time to time by the diminution in hostility between the main contenders. The armaments race has been accelerated, on the other hand, by the outbreak of war or by the fear of an imminent war, when the search for bigger and better weapons has reached massive proportions and consumed much of the available surplus wealth of the societies concerned. This process, already apparent in the nineteenth century, has been accentuated by several factors in the twentieth century.

The main factor has been the increasing relevance or adaptability of technological innovation to purposes of war. In the decades when the steam engine was the dominant power technology, it was not immediately apparent how this somewhat cumbersome and vulnerable prime mover could be adapted to warlike applications. The steam engine, indeed, must be reckoned as one of the most beneficient of human inventions, as it never had more than a supporting role to armed land forces. Admittedly, it did have a big impact on warfare at sea, because it transformed the traditional forms of combat between heavily armed sailing ships into a more mobile form of engagement between screw-propelled steamships. Paddle-wheeled ships had always been too vulnerable to make them attractive as battleships to the navies of the world, but once the screw had taken over as the favoured form of steam propulsion, the switch to steam battleships was swiftly effected. In conjunction with the change to iron and steel construction, and with the introduction of heavy artillery installed in mobile turrets and firing shells instead of the traditional cannon-balls, the transformation to the 'Dreadnought' – a type of large steel ship powered at high speeds by steam turbines and equipped with heavy armour-plating and formidable fire-power – was completed by the years before the First World War. When that war broke out, the leading navies of Europe had established a precarious balance and both sides were anxious to avoid a showdown so that, paradoxically, naval power played only a small part in the hostilities. And where it did become significant, in the submarine war which almost starved Britain to defeat, it was the internal combustion engine and electricity which provided the motive power.

The most important contribution of technology to nineteenth-century warfare, however, was not so much in prime movers as in chemical processes, and particularly in the development of high explosives. These could be manufactured in forms suitable for cartridges, encouraging the evolution of the repeating rifle and the machine-gun, and also in the form of shells which could be fired from very large guns. Such guns could be used in both sea and land war, and they made their mark in the devastating bombardments which preceded any attempted advance in the trench-bound war which pulverized Flanders between 1914 and 1918. The production of explosives for these military engagements became a colossal industry in all the European nations, and justified the description of the First World War as a 'chemists' war'. Of course, it was much more than that, but the heavy reliance on chemical production was certainly a major technological theme of the conflict.[5]

The other important innovations of the First World War derived from the internal combustion engine. They were the tank, a heavy armed vehicle equipped with caterpillar tracks to enable it to cross difficult terrain, and the aeroplane which, as we have seen, had arrived just in time to be adapted to military use and which quickly established a new theatre of war – the war in the air. The larger aircraft were available for bombing, although bomb loads were low and the efficiency of such aerial attack was very limited. Much more effective were the roles of the aeroplane as a reconnaissance platform and as a means of engaging enemy formations in aerial 'dogfights', which were able to endow even the ghastly conditions of the Western Front with a trace of glamour. Meanwhile, behind the lines, trucks and motor cars made a useful contribution to the war, supplementing the railways in bringing up supplies and reinforcements, and in providing ambulance services. The war acted as a powerful catalyst, stimulating production of these new machines and thus helping to create great industries in place of the mainly small-scale and fragmentary production units which had been the normal pattern before.

If the characterization of the First World War as a 'chemists' war' is acceptable, the Second World War can be regarded as a 'physicists' war'. Whereas the technology of the first conflict had been dominated by the production of high explosives, the course of the second was shaped by radar, an electronic scanning technique which gave a vital advantage to the RAF in the Battle of Britain, and by the sophisticated coding and code-breaking machines which made available vital intelligence and prepared the way for the electronic computer. It was ultimately won in the Far East by the success of the physicists in developing an atomic bomb which put Armageddon at the disposal of the Allies. These inventions were derived from a very arcane theoretical physics which was hardly known before the war, but to convert them into engineering hardware for use in radar stations and aircraft required the dedication of enormous resources of capital and skilled manpower. But, however dominant, these innovations were only part of the rapid technological development brought to maturity by the war; others were the jet engine, the helicopter, the V1 and V2 'vengeance' weapons – and particularly the latter, the rocket bomb, which heralded the adoption of a revolutionary new technique which was to open up the frontiers of space in the decades after the war. Add to these the acceleration in the production of antibiotics such as penicillin and the development of insecticides such as DDT, amongst a range of other new techniques, and it can be seen that the Second World War was remarkably rich in technological inventiveness.

An important reason for the wealth of technological innovation in the Second World War was the active involvement of the belligerent states in the promotion of 'research and development'. For the first time, war made innovation an urgent matter of state and, driven by this technological imperative, the nations strove with each other to innovate or perish. The anxiety of the Allies that Hitler could be working towards the production of an atomic bomb was a powerful spur to action, and justified the colossal expenditure on the Manhattan Project whereby the Allies undertook their own programme of atomic development. Such high

levels of commitment could not have been maintained subsequently but for the Cold War between the USA and the USSR,
which obliged the two superstates to prolong their heavy investment of resources in the research and development of ever more
sophisticated weapons. The lesson of modern war has thus been
that no nation can afford to neglect technology any more, and
that so long as a high level of rivalry persists amongst nations it
will remain necessary for them to commit large resources to
weapon development, with significant economic and political
implications for the whole community.

13. The Technological Dilemma

Such has been the awesome size and scope of the impact of technology on modern society that historians have drawn freely on images from myth and legend to describe it. A favourite metaphor has been the myth of Prometheus, stealing fire from heaven for mankind and being punished for his presumption throughout eternity – but fire was not taken away from mankind. Others have been the legend of King Midas, whose wealth-creating touch turned all to gold, and of the sorcerer's apprentice who accidentally unleashed uncontrollable forces. The sword of Damocles, poised in judgement over a divided world, and Frankenstein's man-made monster are other potent myths. In this chapter we are using the image of the paradoxical dilemma in which the subject – in this case, the whole modern world – is presented with irreconcilable alternative prospects which are so intimately related that a clear-cut decision between them is difficult, if not impossible to make. The victim is thus impaled on the horns of the dilemma. It is the technological dilemma because technology provides the matrix of the relationship between the available alternatives.

The nature of the technological dilemma can be succinctly stated. The modern world has become utterly dependent upon technology in many ways, but especially as the means of generating wealth to maintain living standards. At the same time, however, this technology threatens to destroy society. The dilemma is thus that of avoiding technological destruction while retaining the benefits of technological dependence. We have already had occasion to note several aspects of both the dependency on and dangers of technology in the modern world. In particular, our enjoyment of the amenities which sustain a high standard of living in Western nations and the potential for horrific conflict through the use of high-technology weaponry have been described and require no further discussion in this context. For the benefit of clarity, however, it will be useful to review in slightly more detail the nature of the two horns of the technological dilemma before going on to suggest ways in which the dilemma might be resolved.

The sense of increasing reliance on science and technology was well encapsulated by Derek de Solla Price in the title of his stimulating book, *Little Science, Big Science*.[1] In the course of the last two centuries, we have moved steadily from a situation in which scientific discovery and technological innovation were still essentially amateur and small-scale activities, however much they might be encouraged by the social conditions in which they occurred, into a situation in which they are very large and highly organized activities receiving substantial social resources and staffed by an army of professional experts. Price employed some ingenious statistical arguments to demonstrate this transformation from 'little science' to 'big science', and derived from them some fascinating inferences about the implications of the change for modern society. For our purposes it is enough to observe that we have participated in the creation of a world which is dominated by the technology of 'big science', first, by relying upon it for our means of self-defence in an international community which is still divided between nation states which are only too frequently bitterly hostile to each other, and second, by accepting it as the necessary underpinning of the relatively high living standards of the Western nations. Dependence upon technology is now so complete that life without it, if not inconceivable, is certainly very unattractive, and it is doubtful whether the constantly increasing population of the world could be supported without continuing heavy reliance on the materials, instruments and processes of modern technology.

While 'big science' technology has provided a cornucopia of material wealth for those lucky enough to enjoy it in the Western nations, and has underwritten the vital subsistence of the great majority of the world population who do not enjoy these benefits, it has also created problems which pose a serious threat to the very existence of our society, and this is the other horn of the technological dilemma. First, the successes of technology in medicine, pharmaceutics and associated skills have created the conditions within which the staggering population explosion of the last century has occurred, and which continues today without significant abatement. The fact that technology has also provided means of population control through mechanical and chemical forms of contraception has so far made little impact where it is most needed, for even though the developed countries have long been operating a perceptible check on their rate of population increase, powerful cultural and religious prejudices discourage the acceptance of similar constraints in the developing countries, with potentially disastrous consequences for the world economy.

Secondly, the possibility of deploying the instruments of technological warfare in massive self-destruction has loomed over the human species ever since the dropping of the first atomic-fission bombs on Hiroshima and Nagasaki in August 1945. The accumulation of these and the even more devastating hydrogen-fusion devices in the world's armouries makes the extinction of mankind by a self-inflicted orgy of destruction a real feasibility. That is without taking into account the formidable stockpiles of chemical and biological weapons such as poison gas and lethal bacteria, and the so-called 'conventional' means of warfare, all of which

238

are capable of inflicting fearsome and potentially irreversible destruction. Nor does it allow for the potential danger presented by the spread of radiation from nuclear-powered electricity generating stations should the security of these ever be breached in the course of war, nor for the dreadful damage which we know can be caused by an accident in one of these power stations. The nuclear power industry began in the 1950s, full of promise for a solution to the long-term energy crisis as fossil fuels are depleted, but after a series of catastrophic accidents, culminating in that at the Chernobyl power station in Russia, public confidence in the safety of power from nuclear fission has been severely shaken. The continuing unsolved problem of the safe disposal of waste from these power stations has added to public anxiety, and it is now doubtful whether confidence can be restored until techniques can be found for harnessing the power of nuclear fusion, and that still seems to be some decades away. Meanwhile, the existence of nuclear bombs and nuclear power stations poses a threat of destruction and radioactive contamination, and the fact that we have survived this threat for half a century in no way reduces its immediacy or its seriousness.

The threats of population explosion and nuclear extinction are the two most serious legacies of technological revolution with which we have to live, but they are not the only ones. There is also a wide range of environmental problems which have emerged as a result of the application of new technologies since the Second World War. One of the first persons to alert public opinion to these anxieties was the American scientist Rachel Carson, who published her seminal book, *Silent Spring*, in 1962.[2] The book vividly described the insidious effects of new chemical agents in blighting life in town and countryside. At the time, the author was widely condemned by fellow scientists as a scaremonger for suggesting that the indiscriminate use of insecticides such as DDT was anything other than a marvellous benefit and for shaking public confidence in their use. However, the fact that they were poisoning basic food-chains necessary for the sustenance of many forms of plant and animal life, and possibly exerting a malevolent effect on human life also, was begrudgingly accepted over the next few years, and Rachel Carson has come to be regarded as a prophet of ecological consciousness; tight controls have since been placed on the dangerous chemicals to which she drew attention.

Once alerted to this sort of problem, in which some form of environmental pollution disrupts the sensitive ecological balance necessary to maintain the health of specific life-forms, many more were recognized. DDT and other insecticides were being used with dramatic success to boost agricultural production, especially in the developing countries where the 'Green Revolution' had introduced high-yield rice crops, which were more prone to infestation than the older and more robust seed, so that insecticides could not simply be abandoned. Substitutes had to be found and tested for side-effects, and it is now generally realized that any artificial material introduced into the food-chain has to be monitored very carefully. The same applies to artificial fertilizers, used in great quantities in modern agriculture and indispensable to the maintenance of high levels of food production, even though the consequences in terms of run-off of nitrates into rivers and eventually into water supplies poses a difficult environmental and health problem. And the pharmaceutical industry, supplying a stream of new drugs to the market, faces a similar problem in an acute form as every innovation has to be tested rigorously for undesirable side-effects. Even then, no such testing system is foolproof, as the disaster with thalidomide demonstrated in the late 1950s. In this case, a drug with excellent effects as a sedative was found too late to attack the foetuses of unborn children and cause deformities. New techniques can thus be dangerous in quite unexpected ways, and our growing ecological consciousness is providing a valuable warning system against such threats.

Amongst other dangers, the new ecological consciousness has aroused awareness of the problems resulting from the over-exploitation of the natural resources of the world. Soil erosion, for example, has been a fact of life since the earliest agricultural societies, and techniques for avoiding it, such as contour farming and irrigation, are amongst the oldest skills of farming communities. But the pressure of population on existing resources of land in places like subtropical Africa has caused over-grazing of pasture, the over-cultivation of arid land, and the destruction of forest and shrub cover, which are in turn resulting in the massive loss of usable land by wind erosion. Similarly, the pressure of more mouths to feed, in combination with improved techniques, has led to serious overfishing of the oceans, with the consequent depletion of stock. Nylon netting has permitted much more extensive trawling operations than those which were possible with conventional twines, and the indiscriminate use of such techniques has proved harmful to many other species as well as those being sought by the fishermen. A particularly distressing aspect of overfishing in recent years has been the depletion of the world whale population. Whales, the largest terrestrial animals, have been hunted systematically and ruthlessly for their flesh and their oil so that, despite repeated international attempts to control these operations, several species are on the brink of extinction.

Another range of environmental hazards is associated with the pollution caused by burning fossil fuels. The now almost unbelievable filth caused by the mass-consumption of coal in open fires has been brought under control since the Second World War by legislative measures such as the British Clean Air Acts, and persuading people to use oil fuels and electricity rather than coal.

However, the problem has taken a more general and persistent form, which became apparent when the connection between sulphur emissions from large power stations burning coal or oil and the phenomenon of acid rain – which causes the blight of forests and the depletion of fish stocks in rivers and lakes – was recognized by environmentalists. It is possible that a similar ecological relationship exists between car exhausts and some forms of vegetation, causing die-back in neighbouring woodlands, but the mechanics of this are more obscure than those of acid rain. Certainly, however, atmospheric lead concentrations in areas near to motorways and other heavy road transport pose a danger to public health and have promoted the adoption of unleaded petrol. It has become abundantly clear that essential technologies of power generation and transport have been producing excessive amounts of harmful pollutants, even after the more obvious offences of black smoke have been eliminated.

Yet more sinister is the linkage now established between carbon dioxide emissions from burning fossil fuels and what has become known as 'global warming' or 'the greenhouse effect'. It seems that two centuries of heavy consumption of coal, peat and oil fuels has significantly raised the carbon dioxide level in the atmosphere and that this has increased the capacity of the atmosphere to retain heat received from the sun – hence the 'greenhouse' analogy. This effect has been compounded by a different technological side-effect in the form of the depletion of the ozone layer around the earth, caused by the release of chlorofluorocarbons – chloride-based gases used in refrigerators and as compressants in cans for hair-sprays. Large holes have recently been identified in the ozone layer, permitting more of the potentially harmful ultraviolet radiation to penetrate to the surface of the earth. The net result of these processes appears to be a distinct rise in the overall temperature, which has serious implications for climatic changes, land fertility, crop productivity and all that stems from these. Amongst the more disturbing consequences is the likelihood that prolonged warming will contribute towards the melting of the polar ice-caps and that this will cause sea levels to rise and put at risk all human settlements close to present sea levels.

The environmental ramifications of technological development are thus very serious, and the fact that so many of these problems have only recently been identified and are still not completely formulated reflects a certain resistance on the part of world opinion to come to grips with them. It is easier, moreover, to see the offence caused by others than it is to recognize it as one's own. The developed nations have, at least, begun to consider at the highest political levels the implications of these problems for public policy, but the developing nations are too concerned with the more basic problems of feeding a growing number of mouths and paying off interest on capital borrowed from Western banks to give much attention to the less immediately pressing problems of ecological balance. Thus the developed world has watched with increasing alarm as Brazil has been burning off its equatorial forest, knowing that this forest performs an irreplaceable role in maintaining atmospheric oxygen and in moderating climatic changes. But from the point of view of the Brazilian peasant with a large family to sustain – and sometimes with funds from Western bankers to help him – land won from the forest holds out the promise of a livelihood, so he goes on burning it. The British scientist James Lovelock has observed that the envelope of the earth's biosphere has an enormous capacity for correcting automatically any disturbances in the balance of gases in the atmosphere and materials in the oceans, and he has given this self-correcting mechanism the name 'Gaia', after the Greek Earth Mother. But the evidence of modern technological development shows that we are pressing hard on this capacity. The human species, by design or folly, is using technological tools to abuse Gaia.[3]

With his dependence on technology on the one hand, and, on the other, the battery of threats menacing him through his use or abuse of technology, modern man appears to be impaled very firmly on the twin horns of the technological dilemma. The lessons of recent history provide no clear indication either of the capacity or the will of people to release themselves from this dilemma, but it is worth examining such clues as are available in order to make an assessment of the situation. The evidence of some of the more perceptive and eloquent commentators is not immediately encouraging. H. G. Wells, for instance, a pioneering author of science-fiction projections who made some rational but startling forecasts about probable developments, was regarded as the prophet of a new scientific and technological Utopia. But the film made of his story *The Shape of Things to Come* envisaged a horrific period of warfare and world epidemics in the 1940s, and Wells lived long enough to reach complete disillusionment with the idea of progress when he wrote *Mind at the End of its Tether* at the end of the Second World War. Aldous Huxley made similarly gloomy projections of future developments in *Brave New World*, with its strict social distinctions based on genetically engineered biological differences. And Jacques Ellul, survivor of the French Resistance Movement in the Second World War, took an equally sombre view when he wrote *The Technological Society* in the 1950s. To Ellul, technological society was a juggernaut, sup-pressing personality under an efficient centralized bureaucracy: in a sentence, 'Man is to be smoothed out, like a pair of pants under a steam iron.'[4] Ellul did at least present a glimmer of hope for mankind in the possibility of a revival of Christian spirituality. But for George Orwell, in *Nineteen Eighty-four* (1949), no hope is offered in the dark nightmare of human beings suffering eternally under dictatorial regimes equipped with the technological tools of repression. Extrapolating from the totalitarian regimes of the Second World War and after, Orwell could only see the prospect of 'Big Brother', with all the technological advantages in his favour, grinding the face of human individualism – for ever.[5]

Many lesser writers have adopted a similarly gloomy view of the oppressive power made available by technology, and some have taken this view to its ultimate conclusion of denying the possibility of any human influence on the course of a predetermined technological domination. Such complete technological determinism is rare, but the case for it should not be lightly dismissed. At the very least it is necessary to recognize the existence of what Tom Hughes has called 'technological momentum' – the tendency built into every technological system, be it an industry, a power network or a national defence system, to carry on in the direction it was programmed to take at the outset and from which it can sometimes only be deflected with great difficulty.[6] Although there is a measure of determinism about this view, it is not technological determinism, because it assumes the existence of an architect or programmer who has decided what the system should do and who retains certain powers to direct or control this process. We are back, in short, to the idea of technology as a set of tools, to be used or misused by mankind. The fact that these tools have become extremely complex systems makes their control more difficult and the possibility of misuse more serious than with simpler tools, but the essential nature of the relationship remains the same. The creative initiative is still with human beings, and it is up to them to exercise control over the tools at their disposal. To operate control effectively they only need to know where they wish to go. Bacon's vision of acquiring dominion over nature has been largely achieved by means of technology, but human beings need now to think about what they want to achieve by it.

There is, then, a more positive response to the technological dilemma, which is that the worst consequences of the dilemma can be avoided by asserting control and direction over the complex systems of technological society, and as the alternatives are so unattractive it would be folly not to explore this possibility. A number of experiences, historical and contemporary, provide grounds for hope that this approach will work. For one thing, it is clear in retrospect that the decisions of individuals have been important in the processes of technological revolution. It is true that these decisions have stemmed from a great range of motives and most of them have carried very little weight. For millions the choice has always been limited: the harsh pressures of starvation, factory discipline and wage slavery, as well as even more direct political compulsions, have deprived them of significant freedom in this respect. And those who made the key decisions and innovations in the eighteenth and nineteenth centuries did so from a mixture of motives, including financial gain, the love of power and the wish to improve the lot of mankind. The fact that the consequences of these decisions have not always been as intended by those who initiated them does not reduce the impact of the decisions. They remain significant as indications of the ability of individuals to influence the course of events.

Whatever the dominant motive in any particular case, however, the most effective decisions were taken by those merchants and manufacturers, politicians and administrators, who were sustained by a strong sense of self-discipline, which often had roots in a conviction deriving from a clarity about motives and objectives and a resolution to fulfil possibilities which characterize the great missionary traditions of humanity. Such single-mindedness animated the monastic communities which developed agriculture and industries in the wildernesses of medieval Europe, and it supplied the impulse to make the entrepreneurs of early industrialization dedicate themselves to hard work and forgo the enjoyment of quick rewards for their work. When allied to dogmatism and fanaticism, such commitment has been responsible for some hideous tyrannies and abuses of power, but in its creative form it has been capable of remoulding and transforming society. It is probably significant that motivation is weaker now than it was in the early stages of industrialization. There is, to put it in dangerously simplistic terms, less readiness to make sacrifices in the hope of future gains, to plough back profits and to think of long-term objectives rather than short-term comforts. If this analysis is even partially correct, it reinforces the case for strengthening the human motivation as the first step towards acquiring proper control over the course of technological development.

While this motivational factor is of paramount importance in achieving control over technological development, there is plenty of other evidence that the response of modern society to the technological dilemma has not been merely passive and submissive. However much scholars may argue about determinism in theory, in practice we have acquired an elaborate apparatus of instruments for directing technology. We have already observed the success of ecological consciousness in providing an early-warning system in this respect. A great deal of thought has also been expended, especially in the American literature, on 'technology assessment' as a means of controlling the operation of technology. Technology assessment is a technique for the close analysis of historical events which provide possible precedents for contemporary problems. David Collingridge has compared our present ignorance about the future of the microelectronic industry with the situation at the beginning of the present century in relation to the motor car, when, in 1908, a British Royal Commission 'saw the most serious problem of this infant technology to be the dust thrown up from untarred roads', the more profound social consequences of the automobile being then beyond the scope of reliable prediction.[7] Collingridge sees little hope of increasing this predictive power significantly, and is consequently critical of the attempt of the advocates of technology assessment for aspiring to do so. But the aim of learning from historical experience should not be dismissed as worthless, and the hope of using it to make more mature judgements in the present is intrinsically promising.

Other efforts to establish some control over technological development are represented by the terms 'alternative technology' and 'appropriate technology'. Alternative technology is concerned with devising ways of avoiding reliance on fossil fuels and other non-replaceable materials, such as rare metals, by developing alternative techniques and machines. Appropriate technology aims at popularizing techniques which are appropriate to a small-scale, peasant community, with applications such as improved ploughs, bicycle-driven machinery, and small water-wheels for use in fast-flowing mountain streams. Its outstanding spokesman was E. Schumacher, who coined the phrase *Small is Beautiful* as the title of his pioneering book on the subject.[8] Schumacher took the view that Western technology had tended to become gargantuan in its emphasis on size and efficiency, and that giant industrial enterprises on this model had only limited relevance to the needs of peasant societies in developing countries such as India. In their place he advocated a multitude of small enterprises, appropriate to the scale of peasant communities and employing small-scale, or 'intermediate', technology. His views have been widely adopted, and there is a real hope that a degree of control can thereby be fashioned over new technology as it is introduced into the developing countries.

It is far from certain that any of these control mechanisms for determining the direction of technological society can provide anything more than superficial influence over the major technological systems which dominate the modern world. The attempts of the Club of Rome in the 1970s to use computer analyses to predict future technological developments were subject to searching criticism and it was demonstrated that the predictions were in fact largely determined by the programme fed into the computer at the outset.[9] The recognition of the fallibility of this method of control has caused a certain loss of confidence in other methods also, and this brings us back once again to the importance of the human factor in establishing an effective motivation for control. Humankind has gained dominion over nature through technology: so what do we wish to use it for? The avoidance of self-destruction, the stabilization of population, the equalization of wealth and the pursuit of knowledge: a determined effort to achieve these objectives would go far towards resolving the problem of the technological dilemma.

The avoidance of self-destruction must have top priority on any list of objectives, for unless this can be achieved the other targets become irrelevant. The avoidance of a catastrophic war is essentially a political problem, but its solution subsumes the effective control of those technological systems involved in the matrix of national defences and rival capacities for nuclear retaliation. In the last resort, the only guarantee of relative security, which is as much as we can humanly hope for, is the creation of a world community in which the idea of a military attack upon another member is not only unnecessary but virtually impossible because the technological means of making such an attack are subject to an overriding political authority. Such a community would need to be a much larger and more complex state organization than any achieved hitherto, but it remains the rational and logical consequence of any resolution to avoid self-destruction by controlling technology, for if we are to survive as a world community the apparatus of a world state must be acquired sooner rather than later.

When the political structures of survival in a technological civilization are recognized and accepted, it will become possible to apply the resources of the world more effectively than at present to the resolution of other aspects of the technological dilemma, and especially to those calling for social and economic controls in order to achieve measures of population stabilization and wealth equalization. These are political rather than technological objectives, but again technology provides many of the tools by which they can be achieved if the motivation to do so is present. The problems are formidable, requiring the application of enormous resources and consistent commitment, and the substantial and rapid modification of human mores, which, in many parts of the developing world, still place emphasis on unlimited families and the segregation of communities according to their ethnic origin or wealth endowment. It is too much to expect all developing countries to adopt the draconian family-limitation policy applied over the past decade by the government of the People's Republic of China, with its insistence on parents having only one child. But it is a sobering thought that even if this policy could be generally enforced in China, it would still be twenty years before there would be a significant fall in the rate of population increase in that country, which already accounts for one quarter of the human race. Similarly, it would be impossible to introduce overnight policies to break down class and ethnic barriers by measures of social levelling which would remove the more ludicrous imbalances between the rich and the poor of the world. These policies would involve a concerted switch of wealth from the rich countries of the 'north' to the poor countries of the 'south', to adopt the topographically confusing terminology of the Brandt Report, which in 1980 argued strongly in favour of such an objective.[10] Such a dedicated commitment would require the generation of an exceptional degree of societal motivation.

To state thus the measures necessary to escape from the technological dilemma shows the stark enormity of the problems confronting modern technological civilization. It would be futile, however, to attempt to dodge the issues: these are the problems which have been created by the processes of technological revolution and now, at the end of the twentieth century, we cannot avoid attempting to solve them, if only because the alternatives are so bleak. But it is implicit in this account that solutions are available, and it is in this optimistic spirit that the problems are confronted. In a sense, the crucial factor comes from within the fabric of the technological system. This is the motivation which derives from the pursuit of knowledge, the mainspring of scientific discovery and technological innovation. It is this objective, raised to the level of a human mission to explore the universe, which provides a key to the logistics of survival. It will be the subject of our final chapter.

14. The Way to the Stars

The idea of progress has not been as popular in the twentieth century as it was in the previous hundred years. Up to the First World War there was a widespread assumption that the world was getting steadily better. The manifold successes of industrialization in increasing productivity and raising standards of living, in conjunction with the apparent advances in liberal democracy in many countries and the euphoria associated with national expansion as the European and North American states asserted their technological superiority over less developed parts of the world, ensured that the notion of material and moral progress towards a higher plane of human achievement was virtually taken for granted. It found many expressions in the literary culture of the nineteenth century, and was powerfully reinforced by the scientific adoption of evolution as a convincing explanatory theory. Human history could be persuasively expressed in terms of social Darwinism, in which the fittest – that is, the best – successfully survived the conflicts of life. The fact that this was not exactly the meaning which the cautious and scrupulous scientist Charles Darwin had meant to convey was irrelevant: his theory of evolution by natural selection provided a potent metaphor for the inevitability of human progress. Popular opinion in the West found the idea of progress intellectually satisfying, and even the dynamic historical materialism of Karl Marx derived much of its appeal from its resonance with this general assumption of the inevitability of progress.[1]

This easy optimism met its nemesis in the horrors of trench warfare in the First World War. The colossal loss of life, and futility of the conflict, made sure that progress would no longer be regarded as something which could be taken for granted. Subsequent twentieth-century experience of demonic totalitarian regimes, the holocaust of Jews and innocent civilians in the Second World War, and the ever-present threat of nuclear annihilation which has menaced the world since 1945, have tended to confirm a cynical attitude towards progress. The human species no longer appears to enjoy unchallenged moral superiority over other life forms but seems, rather, to be a species which, at odds with all others, is capable of killing and destroying for pleasure. The moral high ground occupied with such eloquence by nineteenth-century political and religious leaders, which enabled them to speak of the 'white man's burden' and the civilizing qualities of Western culture, has now been deserted. It is no longer possible to believe that all things will necessarily work together for the good and benefit of humankind.

Nevertheless, some of the content of the idea of progress deserves to be salvaged. Despite all the disillusionments of the twentieth century so far as the moral perfectibility of human beings is concerned, there remain many impressive material achievements, which constitute progress of a sort. And in so far as these material achievements have maintained the increase in productivity established by the early successes of industrialization, and have vastly improved the quality of life available to most people living in Western societies, they are not to be dismissed as insignificant. Indeed, they represent a very substantial advance in the ability of people to make and do things, and as such they demonstrate that the conditions of technological revolution are still exceedingly active. In other words, more than ever before, the tools of technology are available to enable human societies to tackle the endemic problems of poverty, inadequate food production, social inequalities and such like. As we have already observed, it is the nature of the technological dilemma that technology has made itself indispensable while at the same time posing the ultimate threat to human existence. But the continuing pace of technological progress has at least opened up the possibility of escaping from the consequences of the dilemma. Once again, it is primarily a question of human motivation: of resolving whether or not the present generation of human beings is capable of responding to the challenge with a sense of mission. Progress, therefore, although real in material terms, has been partial, and requires active human commitment to become generally effective.

Throughout this book we have been concerned with exploring the processes of technological revolution whereby Western societies have been continuously transformed during recent centuries. The transformation has been one of technological revolution in the sense that, despite all its enormous complexity, the most pervasive and persistent stimulus to change has been the technological combination of scientific discovery and technical innovation. It is not denying the existence of other powerful social forces such as popular pressure, class conflict and nationalism to place this emphasis on technology. But it does involve looking at the development of the modern world in a particular way, and one which provides a more convincing explanation of the pattern of these developments than any of the more traditional historical treatments.

The unique quality of the present situation has appeared at every stage of our analysis. Never before have the conditions of technological dilemma, described in the last chapter, existed. After millennia of evolution, human society has achieved in the twentieth century both an unprecedented dependence on technology and an extraordinary capacity for self-destruction. The situation is unique to Western civilization, because in no previous human society have the processes of technological revolution been able to operate so comprehensively. In no other society has the capacity for massive wealth creation through manufacturing processes been mastered. Nor are there any historical precedents for the achievements of the West in transport and communications. The fact that we have come to take for granted in the twentieth century the facilities for flight, space travel and instantaneous communication throughout the world, demonstrates the astonishing novelty of contemporary experience and indicates the need for extreme caution in deriving historical lessons and projections. The world has entered new territories of the psyche and the intellect as well as of material conditions, and many of the traditional grounds for historical judgement, such as seeking appropriate precedents, may no longer be valid. Certain parameters, however, can still be observed.

Because technological revolution has taken place in a particular society over a definable period of time, it has been possible to distinguish clear limits to our study. The area covered, however, remains vast, and there is no sense in which this can be regarded as an exhaustive examination. It is, rather, an outline sketch or survey, picking out the key features and trying to establish the relationships between them. The inadequacies of such a treatment are self-apparent, but in the circumstances it is more important to establish the general nature of the landscape than to dwell at length on any details. And even if no clear pattern emerges from the examination, there are at least some dominant themes which suggest lines of interpretation. Thus, in the course of surveying the main aspects of technological development – the sources of power, the application of this power to manufacturing industry, transport, communications and infrastructure, and the impact of technological change on people collectively and as individuals – we have repeatedly had occasion to observe the intimate relationship between technological change and the social conditions in which it occurs.

In particular, we have stressed the importance of an environment receptive to technical innovation; we have identified the characteristics of such an environment as comprising a 'package' of relative liberalism, tolerance and encouragement of individuality; and we have emphasized that the moment of technical invention is essentially one of human creativity, being as such in the last resort unpredictable and uncontrollable. Once an invention has occurred, however, we have observed time and again that the processes of innovation and development by which it is converted into a successful technique are profoundly conditioned by the environment, and that success depends as much on the existence of economic resources, craft skills and appropriate rewards for entrepreneurship, as it does upon technical excellence. Moreover, we have been especially intrigued by the 'ratchet' mechanism which frequently appears to operate between concurrent technical developments, so that a successful innovation in one field can provide a useful impulse to a different development. For instance, the way in which the mature technology of steam power provided a secure basis for the new technology of internal combustion, or the way in which a network of electrical power distribution enabled the electronics industry to develop illustrates this relationship and suggests that a crucial feature of a congenial environment is accessibility to relevant technical experience. It follows from this that the ratchet metaphor does not only apply to the artefacts themselves, even though these demonstrate the more obvious points of engagement between the old and the new forms of development. Equally involved, however, is the social

256

milieu in which the engagement takes place: the education of the artificers concerned, their ability to communicate with each other and the degree to which their communication is controlled by social, radical, legal or political factors.

Although this package of factors comprising a social environment congenial to technical innovation can be broadly defined, we have recognized that its existence is not essential to the use of technology, and that some of the more horrific abuses of modern technology have occurred when brutal dictatorships have appropriated the technical means to their own malevolent ends. This led us on to speculate about the necessity for some form of world community with the powers of a world state which would monitor the occurrence of such abuses and deploy resources to prevent their repetition. We held out the hope that, by bringing such a measure of collective self-protection into the international order, it would be possible to tackle the problems of population explosion and inequalities in wealth which are also threats to the stability of the world. But we recognized that, just as technical invention depends, in the last resort, on the human individual, so too does any concerted confrontation of the technological dilemma. And the necessary factor here, tantalizing in its uncertainty, is that of human motivation. Once again, we pose the question: having substantially achieved the Baconian objective of dominion over nature, what do we want to use it for?

There are many possible answers to this question, but they fall into two general categories: those which opt for short-term enjoyment or hedonistic objectives, and those which emphasize long-term objectives, with the implication of readiness to forgo immediate gains in the interest of some greater eventual benefit. The short-term view tends to reinforce the pessimistic prognostications of the technological determinists, because those who hold it take advantage of the opportunities for self-indulgence made available by the technological revolution without attempting to direct it. Thus, in the absence of controlling principles, the momentum of technological development acquires a logic of its own. On the other hand, a long-term view would provide specific goals for technological development and attempt to control technology in accordance with these goals. Such goals require dedication and commitment on the part of those concerned. They may also be illusory, as many political programmes for Utopian communities have been. Motivation of this sort does not come easily, because it can only be sustained by confidence in the future and by a degree of self-denial and self-discipline in the present. It is, however, of the utmost importance if we are to find a satisfactory way out of the impasse of the technological dilemma, and needs to be encouraged. In particular, a sense of technological mission is desirable in order to face the future with real hope.

Technological mission involves confidence in human rationality and creativity. It aims at directing the resources of technological revolution towards the solution of social problems such as the population explosion and wealth inequalities, but it does so on the assumption that man exists to pursue knowledge and the wisdom which comes from this pursuit. Humankind's searches for knowledge of itself and of the universe is an infinite task so that, provided we can solve the immediately pressing problems of the technological dilemma, we can undertake our destiny as the species capable of acquiring and using new knowledge. Important projects currently receiving attention include: the human genome project, to map the genetic structure of human life; the exploration of microelectronics and laser technologies; the promotion of the Green Revolution, to improve the productivity of crops and animals and to discover ways of harvesting the deserts and the oceanic floor of the continental shelves; the search for controlled nuclear fusion and other non-pollutant sources of energy – the list is potentially endless. Beyond them all, it is the destiny of mankind to explore the universe. This is the greatest of all the objectives of technological mission, both in its scale and in its potential for enriching human experience.

Any idea of the 'conquest of space' is an illusion, because the space of the observable universe is so vast that no such puny species as *Homo sapiens* is ever going to be able to conquer it. But it is there to be explored and, by a remarkable synchronization, the means of undertaking such exploration have become available in the second half of the twentieth century just as the technological dilemma has become acute. Even before the means were there, human imagination had pushed beyond the confines of terrestrial life. Jules Verne conceived a gun big enough to achieve escape velocity for a projectile which could then make the journey to the Moon. H. G. Wells suggested the use of an anti-gravity paint to achieve the same objective. We have now abandoned the big-gun solution to the technical problem of a space launch, and the anti-gravity device has yet to be invented, but the development of rocket technology, and especially the multi-stage rocket, has enabled us to cross the threshold to space travel. The rocket stemmed from theoretical work by the Russian Konstantin Tsiolkovsky and some largely abortive experiments in America and Europe by the American Robert H. Goddard and the Romanian-born German Hermann Oberth in the first half of the twentieth century. It was then developed as the V2, the second 'vengeance weapon' of Adolf Hitler, in the Second World War. The brilliant engineering team under Werner von Braun, working at Peenemünde on the Baltic island of Usedom, developed a rocket propelled by burning alcohol and liquid oxygen; the rockets reached a height of over a hundred miles and could deliver a formidable high-explosive warhead on London and other targets in south-east England. The war ended before the full potential of this weapon could be achieved, but the expertise (including von Braun, who went to America) was acquired by the United States and Soviet Russia, and provided the basis for the rocket-development programmes in these two superpowers.

So was the Space Age born and the space race begun. In both America and Russia, military technologists experimented with surviving V2s and started to design multi-stage rockets in the decade after the Second World War, while science-fiction authors ranged with astonishing inventiveness over the possibilities opened up by the new technology and Arthur C. Clarke, in Britain, envisaged the technical details of a satellite communication network.[2] Then, on 4 October 1957, to the consternation of Americans, the Russians launched Sputnik 1, the first successful artificial satellite to be placed in orbit round the Earth. It was only a small sphere with an 83 kg instrument package which broadcast a radio bleep to listeners below, but it signalled Russian mastery of the technique for achieving escape velocity through a rocket burning in two or three successive stages. The United States immediately devoted massive resources to overtaking this Russian initiative, but it took them over a decade to do so, during which time Russian technology remained in the ascendancy.

The space race may be characterized as having passed through four chronological but overlapping phases, in which the Russians won the first two and the Americans the second two. The first phase was concerned with increasing the thrust of rockets to put larger and more complex satellites into Earth orbit, and with exploring their possible uses in communications, in weather observation, in monitoring military information, and in topographical and geological surveying. The second phase was that of manned space flight. It began with the successful orbit of the Earth by the Soviet cosmonaut Yuri Gagarin on 12 April 1961, in the space vehicle Vostok 1. His flight demonstrated mastery of the complex problems of weightlessness and of safe re-entry into the atmosphere. It was followed by a series of Soviet and United States space flights in which the techniques of space rendezvous and docking were rehearsed. Men were kept in space for up to a fortnight, and made the first 'space walks' outside their craft. Russian space technology has since tended to specialize in the construction of ever-more elaborate space stations, with cosmonauts remaining on duty for up to a year at a time and undertaking constructive operations outside the station.

It was in the third phase, concerned with the exploration of the Moon, that America at last drew ahead of Russian technology. The phase began, however, with another Russian achievement when Lunik 1, launched on 2 January 1959, became the first spacecraft to escape the gravitational field of the Earth, to fly past the Moon and enter an orbit round the Sun. In the process, human beings received their first views of the side of the Moon which is turned permanently away from the Earth, showing a uniform pattern of cratering but with fewer 'maria', dark smooth areas, than on the more familiar side. Following missions hit the Moon, flew round it and, on 3 February 1966, achieved a soft landing. By this time, however, the Americans were catching up, making an excellent photographic reconnaissance of the surface of the Moon with their Ranger and Orbiter programmes, and, on 2 June 1966, made their own soft landing on the Moon with Surveyor 1. Meanwhile, the size and power of launching rockets steadily increased, so that by the late 1960s the giant Saturn V rocket, standing 108 m high on its launching pad, made it possible for the United States to embark upon the Apollo series of spacecraft, with capsules designed to carry three astronauts to the Moon and back. Instead of trying to land in one piece and then return from the lunar surface, an intricate but ingenious manoeuvre was devised whereby only a lunar module landed, from which the upper portion carrying two astronauts blasted off to rejoin the third member of the team who remained in orbit round the Moon in the parent craft. This exercise was brilliantly successful when, on 20 July 1969, Neil Armstrong and Edwin Aldrin climbed out of the lunar module of the Apollo 11 spacecraft to become the first human beings to step upon the surface of the Moon. There were five more successful landings and one aborted attempt (Apollo 13, which narrowly escaped disaster when its fuel tanks blew up in flight) before the programme was abruptly closed down. This was partly an economy measure, in face of the enormous costs of the space missions, but it also reflected the fact that the Soviet space programme made no attempt to follow the Americans in manned exploration of the Moon.

While these phases were still being pursued, the fourth and most open-ended phase of space exploration had already begun, with the use of spacecraft to investigate the other planets of the solar system and beyond. As Earth's closest neighbour amongst the planets, Venus was the first to receive attention. The American Mariner 2 was launched on 27 August 1962, and passed by Venus in the following December. It was followed by several other craft, both Russian and American, some of which went into orbit round Venus or landed on the surface. The combined results of these investigations demonstrated conclusively that Venus possesses an atmosphere which is extremely inhospitable to life as known on Earth, with surface temperatures of 900°F, which means that any probes to the surface can only function for a very short span. Nevertheless, the examination of the surface features of Venus, permanently hidden by cloud from visual observation, has steadily proceeded, and scientific knowledge of the planet and its atmospheric mechanisms has been greatly enlarged.

The exploration of Mars has so far been conducted primarily through the United States series of Mariner and Viking space probes. Soviet probes have been dispatched towards the planet, but have failed to perform satisfactorily. In 1965, the Mariner 4 spacecraft flew past Mars and conducted a preliminary photographic survey, making the surprising discovery that much of its surface closely resembled the heavily cratered appearance of the Moon. This came to seem less astonishing when it was discovered that the surfaces of Mercury and of most of the larger satellites around the giant planets were similarly marked, and it was realized that the early solar system must have been subjected to massive bombardment by interplanetary debris, the evidence of which has been eliminated on those planets with sufficiently active weather systems. On Mars, with a thin atmosphere and correspondingly weak weather activity, the obliteration is partial, but hope remained of finding some form of life until 1976, when two Viking spacecraft landed on the surface to take photographs and to make chemical analyses of soil samples which were designed to detect the presence or remains of organic material. These tests encountered mechanical difficulties, but their results were generally interpreted as negative. Mars, like Venus, is thus probably a 'dead' planet, but it remains the subject of intense scientific interest.

By the 1980s, the space race had been virtually abandoned. The Russians had concentrated on a painstaking programme of building up the endurance of their cosmonauts by keeping them for long periods in their Mir space stations. The Americans, also, gave some attention to establishing a permanent space station in orbit round the Earth, but the costs of achieving this on the scale envisaged by some of the more visionary projections became prohibitive, and it was quietly shelved. The American space agency, NASA, which continued to receive substantial support from the American taxpayer, concentrated instead on devising a more economical method than the single expendable rocket of getting a payload into orbit, and devoted its resources to the space

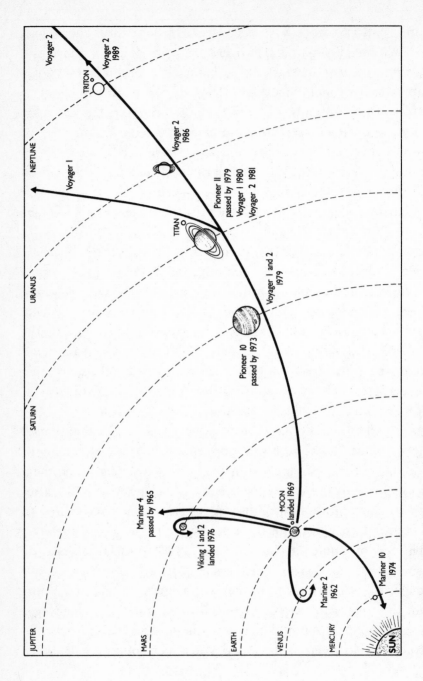

Figure 13. Exploration of the solar system (orbits and planets not to scale)

shuttle. This craft was designed to be launched like a rocket, but then jettisoned its empty fuel tanks before going into orbit and eventually made a controlled descent back into the atmosphere. The shuttle had many teething problems, but eventually seemed to be working smoothly and performing a valuable task as a work-horse ferrying satellites and equipment into space, until the catastrophic accident on 28 January 1986, when the shuttle Challenger blew up shortly after launch, with the loss of her crew of seven. This was a serious blow to American confidence in the space programme, and allowed the more conventional rocket technologies of the Russians, the European Ariane programme, and even the Chinese space programme, opportunities to develop. But it seems certain that the long-term future of space exploration will depend on a viable reusable shuttle technique, so American perseverance with this is likely to be rewarding.

Meanwhile, several encouraging lines of inquiry have been pursued with the aid of space vehicles. The inner planets have been systematically surveyed and mapped, while magnetic fields and the solar wind have been examined. The approach of Halley's comet to the sun in 1986 was intercepted by the space probe Giotto, which passed right through the head of the comet and sent back revealing photographs of its structure. The Hubble space telescope was launched in 1990, and despite early disappointments regarding the accuracy of its main mirror, it promises eventually to send back information about distant celestial objects which cannot be observed from the surface of Earth because so much light radiation is absorbed by the atmosphere. The Russian cosmonauts have maintained a rigorous programme of manned space operations designed to test the adaptability of the human body to prolonged space flight, with the presumed intention of undertaking a voyage to Mars if all the problems can be overcome.

The most impressive of all the preliminary steps as humankind sets out on the way to the stars have been those of the deep-space probes Voyager 1 and 2. These were launched by NASA in 1977 and spent the next twelve years exploring the giant outer planets of the solar system and sending back stunning photographs of these planets and their moons. First, in 1979, came pictures of Jupiter, with its massively disturbed atmosphere, its powerful magnetic field, a thin planetary ring which had not been seen before, and the astonishingly varied collection of moons, from cratered Ganymede through ice-glazed Europa to the volcanic inferno of Io, the innermost moon. Then, in 1980 and 1981, the two Voyagers passed Saturn, providing spectacular evidence about the structure of its ring system and its large flock of moons. Voyager 1 was deliberately deflected out of the plane of the solar system in order to pass by Titan, the largest moon of Saturn, which was found to possess a thick methane atmosphere completely obscuring the surface. Voyager 2 continued outwards along the planetary plane, and in 1986 passed Uranus, observing the thin rings and weirdly contorted moons. In 1989 it passed its last planetary milestone in the shape of Neptune, which was found to be an astonishingly blue planet with an active atmosphere. Its large moon Triton was even more of a surprise, for here, on the very edge of the solar system, was a world displaying a thin atmosphere and active vulcanism on its surface. This was a fitting end to a voyage which had demonstrated technological wizardry and sophistication, and which had transformed our scientific knowledge of the Sun's family of worlds.

It was not, however, the end of the voyage, because both Voyagers 1 and 2 are now travelling out of the solar system, having had their speed boosted by the gravitational field of each of the giant planets as they have swung past them. They carry coded messages for whatever intelligent beings may eventually intercept them in some future time and place as they journey out into interstellar space. But their departure from the solar system represents the end of this phase of space exploration. The preliminaries have now been completed. Our scientific knowledge of the solar system has been hugely enlarged, and fascinating areas opened up for further research. The space race is over, and we must now look forward to closer collaboration between the leading participants in space exploration to ensure the most effective pursuit of the investigation. For out there, on the way to the stars, lie some of the great objectives of the technological mission: clues about the nature of the universe, the structure of matter, the origin of life and the matrix of time itself. And the certainty that, if we pursue the search with sufficient diligence, we will encounter other forms of intelligent life. These amount to a spellbinding and infinite task, and one worth the application of all the resources of mind and spirit of which humankind is capable.

There have been many objections to this programme of space exploration, and it would not be reasonable to regard these as trivial or unimportant. There has been, first, vociferous objection to the cost of space research, although it should be observed that the American taxpayer, whose democratic voice is the only one which has carried any weight, has been amazingly long-suffering about this expenditure. However, the criticism has compelled attention to economies such as the attempt to develop a reusable shuttle in place of the enormously wasteful rocket launch, and at moments of acute financial stringency the programme has been curtailed. The objectors have also observed that the NASA programme has become intimately related with defence expenditure, and in the days when the Strategic Defence Initiative (more popularly known as 'Star Wars') was being actively promoted, the link was very close indeed. The whole space race, after all, was a by-product of the Cold War, and the diminution of this conflict in 1990 implied a lessening of the commitment to space research.

Perhaps most telling of all, however, is the plea that resources devoted to the space programme could be applied with more immediate benefit to relieving world hunger and poverty. Certainly the prosecution of space research is no excuse for not developing humanitarian programmes also. Ideally, both are required, and they support each other because, in the end, the objectives of both are the same: seeking to ensure the survival of the species through greater self-knowledge, acquired through a more profound insight into the nature of the universe.

The processes of technological revolution have thus, despite their great complexity and the challenging problems which they have created in the shape of the technological dilemma, opened up vistas of possible solutions which, if recognized, could provide directions and goals which would enable us to encompass the immediate problems of the modern world and to draw inspiration from a technological mission which points us firmly to the future and to the stars. The responsibilities of our historical situation are tremendous, but our generation is truly privileged to live at this unique cross-roads in human experience when, however great the dangers, the prospects are so great and so hopeful. It is not the business of historians – not even historians of technology – to predict the future, but such is the extraordinary quality of the world situation at the end of the twentieth century, when so much has been achieved in such a short space of time by technological intelligence, that some attempts at projection are justified. What-ever happens in the next few decades, the continued impact of technological revolution is sure to be dramatic, with far-reaching consequences for our planet, the environment and the viability of humankind. In the end, our success or otherwise in achieving the survival of human life will depend upon our readiness to search for knowledge and understanding, and this in turn stems from the spirit with which we approach the immense task of education, the painstaking collection of evidence and the judicious judgement with which we select the valid from the invalid. Like the journey to the stars, the search for self-knowledge is never-ending, but the experience of technological revolution has made both the journey and the search worth while.

Notes and Bibliography

1. The Nature of Technology

In recent decades there have been several large-scale surveys of the history of technology. The most accessible to English-speaking readers is probably that edited by Charles Singer, E. J. Holmyard, A. R. Hall and Trevor I. Williams: *A History of Technology*, published by the Oxford University Press. The initial five volumes, carrying the account down to 1900, came out between 1954 and 1958, and the series was subsequently extended with two further volumes covering the twentieth century, *c*.1900 to *c*.1950, edited by Trevor I. Williams, in 1978. This work remains a very useful compilation, although it is inevitably uneven in quality and is strictly for reference purposes rather than easy reading. Much more manageable is the concise version, *A Short History of Technology*, written by T. K. Derry and Trevor I. Williams, Oxford, 1960, and the sequel covering the last two volumes of the series, *A Short History of Twentieth-Century Technology*, by Trevor I. Williams, Oxford, 1982. Also agreeable is the study edited by Melvin Kranzberg and Carroll Pursell, intended in the first instance for the use of American students: *Technology in Western Civilization*, 2 vols., New York and Oxford, 1967. There are good French general histories which have been made available in translation, especially that by Maurice Daumas, published originally as *Histoire générale des techniques*, 4 vols., Paris, 1962, and the somewhat more idiosyncratic work edited and substantially written by Bertrand Gille: *The History of Techniques*, 2 vols., Switzerland, 1986. Robert J. Forbes, *Man the Maker*, London and New York, 1950, provided a good short history of the subject, and the same Dutch historian contributed to R. J. Forbes and E. J. Dijksterhuis, *A*

History of Science and Technology, 2 vols., Baltimore and Penguin Books, 1963. Friedrich Klemm's useful selection of readings in the history of technology, published first in German, was translated by Dorothea W. Singer as *A History of Western Technology*, London, 1959. Ian McNeil, ed., *An Encyclopaedia of the History of Technology*, London, 1990, is a convenient compilation, and my own article, 'Technology, History of' in *The Encyclopaedia Britannica*, 15th edn, 1975, may serve as a general background. Also, R. A. Buchanan, *History and Industrial Civilization*, London, 1979, raises some of the speculative issues about technology in modern society. Such issues also receive regular attention in the columns of *Technology and Culture*, the quarterly journal of the Society for the History of Technology (published by Chicago University Press), while the *Transactions of the Newcomen Society* have maintained an annual stream of expert studies since 1921. Another periodical series, originally edited by A. Rupert Hall and Norman Smith, is *History of Technology*, which has appeared more or less annually since 1976 and is now at Vol. 12, London, 1990, edited by G. Hollister-Short and Frank A. J. L. James. Also worth mentioning, as the only general bibliographical work on the history of technology, even although it is now somewhat dated, is Eugene S. Ferguson, *Bibliography of the History of Technology*, Cambridge, MA, and London, 1968. There are annual bibliographical updatings in the subject in *Technology and Culture*.

On the archaeological aspects of history, V. Gordon Childe, *What Happened in History* is a classic and is still a stimulating read even though it was written in the 1940s. More recent archaeological speculations about early cultures are well represented by Colin Renfrew, *Archaeology and Language*, Penguin Books, 1987. For the use of physical evidence in modern economic and social history, see R. A. Buchanan, *Industrial Archaeology in Britain*, Penguin Books, 1972, 2nd edn, 1982.

As far as the history of science and its relationships with technology and society are concerned, an excellent introduction is the trilogy by S. Toulmin and J. Goodifield published by Penguin Books, *The Fabric of the Heavens*, 1963, *The Architecture of Matter*, 1962, and *The Discovery of Time*, 1965. Thomas Kuhn, *The Structure of Scientific Revolutions*, Chicago, 1962, made an important contribution to our understanding of this field.

Notes and Bibliography

It is not necessary to agree with all the interpretations of Arnold Toynbee's magisterial work, *A Study of History* (12 vols., 1927–61; illustrated single-volume edition, revised and abridged by the author and Jane Caplan, London, 1972), to appreciate his powerful insights regarding the evolution of human civilizations. On medieval aspects of the subject, Lynn White Jr, *Medieval Technology and Social Change*, Oxford, 1962, is a deeply stimulating and inspiring book, and Jean Gimpel, *The Medieval Machine*, London, 1976, 2nd edn, 1988, is a lively read. Joseph Needham is still engaged on the massive and definitive study, *Science and Civilization in China*, publication of which, by Cambridge University Press, began in 1954 and is proceeding in many volumes and subdivisions with the assistance of various Chinese scholars. Needham's illuminating ideas about the role of the mandarins in social and technical change in China are more accessible in his collection of essays: *The Great Titration – Science and Society in East and West*, London, 1969. There are no comparably synoptic works for the civilizations of Islam and India, but see Donald Hill, *A History of Engineering in Classical and Medieval Times*, London, 1984, for some useful cross-cultural comparisons; and J. Keay, *India Discovered*, London, 1988, for a fascinating account of the discovery of Indian traditions by the West. Two other recent works are of particular interest to the matters raised in this chapter – George Basalla, *The Evolution of Technology*, Cambridge, 1988, and Arnold Pacey, *Technology in World Civilization*, Oxford, 1990. Basalla develops an ingenious model of technological evolution, comparing it with natural evolution, while Pacey presents attractively a cross-cultural survey of technological developments over the last millennium.

2. The Process of Technological Revolution

There have been several attempts to provide historical explanations of modern society which have placed heavy reliance on a technological component, although the interpretation of this role as active or passive, benevolent or malign, has varied greatly. Marx and Engels were amongst the first to recognize its importance, emphasizing its function in accentuating the divergence between the capitalist and proletarian classes.

272

Notes and Bibliography

Many prominent economists have suggested helpful analyses of technological processes as agents of economic growth. Sociologists of many persuasions have been intrigued by the interaction of technology and society. Planners like Lewis Mumford and philosophers such as Jacques Ellul have also had much to say about the role of technology in forming social attitudes. It has thus been an area of significant intellectual speculation from many different points of view in the twentieth century.

1. Edison's remarks are frequently quoted, in several different forms. But see Wyn Wachhorst, *Thomas Alva Edison – an American Myth*, Cambridge, MA, 1981, pp. 13–14, for a particularly entertaining account.
2. Thor Heyerdahl published popular accounts of the voyages which he made to support his theories, *The Kon-Tiki Expedition*, London, 1950, and *The Ra Expedition*, London, 1971.
3. David J. Jeremy, *Transatlantic Industrial Revolution – The Diffusion of Textile Technologies between Britain and America, 1790–1830s*, Cambridge, MA, and Oxford, 1981.
4. Lewis Mumford, *Technics and Civilization*, London, 1934, was the first of several stimulating works from this pen. The fullest statement of Marxist thinking about the economic consequences of technology is in Karl Marx, *Capital*, Vol. 1, London, 1867, and Penguin edition, 1976. S. Lilley, *Men, Machines and History*, 2nd edn, London, 1965, represents a more recent expression of these views. See also Nathan Rosenberg, 'Marx as a Student of Technology', Chapter 2 of *Inside the Black Box – Technology and Economics*, Cambridge, 1982 – 'whether Marx was right or wrong ... his formulation of the problem still deserves to be a starting-point for any serious investigation of technology and its ramifications' (p. 34). And Rosenberg goes on to say: '... Marx is insistent that technology has to be understood as a social process. The history of invention is, most emphatically, not the history of inventors' (p. 48). From the point of view of the present study, the implied antithesis is not valid: technological revolution is *both* social process and the result of individual initiatives.
5. For powerful statements on technology by economists, see J. A. Schumpeter, *Business Cycles*, New York, 2 vols., 1939, and *Capitalism, Socialism, and Democracy*, New York, 1942; and N. Rosen-

berg, ed., *The Economics of Technological Change*, Penguin Books, 1971. Also important is A. P. Usher, *A History of Mechanical Inventions*, 2nd edn, Cambridge, MA, 1954. Counterfactual arguments have been used to evaluate the contribution of railways to the American economy in R. Fogel, *Railroads and American Economic Growth*, Baltimore, 1964, and there have been many other applications of the technique. For a general review of the theory in the history of technology, see Andrew Jamison, 'Technology's Theorists – Conceptions of Innovation in Relation to Science and Technology Policy', in *Technology and Culture*, Vol. 30, No. 3, July 1989, pp. 505–33. See also R. A. Buchanan, 'Theory and Narrative in the History of Technology', in *Technology and Culture*, Vol. 32, No. 2, April 1991, pp. 367–76.

6. See especially the recent work by Wiebe Bijker, Thomas Hughes and Trevor Pinch, eds., *The Social Construction of Technological Systems*, Cambridge, MA, and London, 1987. For a more traditional sociological perspective, see Robert Merton, *Social Theory and Social Structure*, New York, 1957.
7. J. Burke, *Connections*, London, 1978.
8. Arnold Pacey, *Technology and World Civilization*, Oxford, 1990.
9. G. Basalla, *The Evolution of Technology*, Cambridge, 1988.

3. The Ascendancy of Steam Power

The general subject of the development of power has been reviewed by Fred Cottrell in *Energy and Society*, New York, 1955. See also A. R. Ubbelhode, *Man and Energy*, 1954, and Penguin Books, 1963.

On wind and water power, a good general account written from a strongly architectural point of view is that by John Reynolds, *Windmills and Watermills*, London, 1970. Rex Wailes, *The English Windmill*, London, 1954 is still a standard work, although more recent 'molinologists' have added richly to scholarship on this subject. A. P. Usher, *A History of Mechanical Inventions*, 2nd edn, Cambridge, MA, 1954, and A. F. Burstall, *A History of Mechanical Engineering*, London, 1963, are both good general accounts.

There is extensive literature on steam power, to which a recent and able contribution is R. L. Hills, *Power from Steam*, Cambridge, 1989. See also H. W. Dickinson, *A Short History of the Steam Engine*, 1938, reprinted London, 1963; and R. A. Buchanan and George Watkins, *The Industrial Archaeology of the Stationary Steam Engine*, London, 1976. For an economic interpretation of the impact of steam power, see G. N. von Tunzelmann, *Steam Power and British Industrialization to 1860*, Oxford, 1978. L. T. C. Rolt, *Thomas Newcomen – The Prehistory of the Steam Engine*, Dawlish, 1963, revised as L. T. C. Rolt and J. S. Allen, *The Steam Engine of Thomas Newcomen*, Hartington, 1977, provides an excellent introduction to the early history. Carroll W. Pursell Jr, *Early Stationary Steam Engines in America: A Study in the Migration of Technology*, Washington, 1969, gives a vivid insight into the transfer of the technology to the New World, and Eugene S. Ferguson contributes to the same theme with *Oliver Evans – Inventive Genius of the American Industrial Revolution*, Delaware, 1980. Evans's high-pressure steam engine constitutes a 'parallel invention' to that of Richard Trevithick. Another fascinating regional study is that by Svante Lindqvist: *Technology on Trial: The Introduction of Steam Power Technology into Sweden, 1715–1736*, Uppsala, 1984.

1. See Cottrell, op. cit., for his discussion of the 'energy gradient'.
2. Svante Lindqvist, op. cit., gives a detailed account of Triewald's engine.
3. Carroll Pursell, op. cit., provides an account of the involvement of the Hornblowers in America.
4. On patent law and its implications for the history of technology, the recent work by Christine MacLeod, *Inventing the Industrial Revolution: The English Patent System, 1600–1800*, Cambridge, 1988, is very informative.

4. Internal Combustion and Electricity

The main texts for steam technology are as in the previous chapter. Rollo Appleyard, *Charles Parsons*, London, 1933, is disappointing as a biography, but it contains a useful summary of the development of the steam turbine. Coverage of the internal combustion engine is frequently associated with accounts of its application in transport, in either the automobile or the aeroplane. But there is an excellent introductory account by D. C. Field in *The History of Technology*, Oxford, Vol. V, Chapter 8. A useful introduction to electrical power technology is Percy Dunsheath, *A History of Electrical Engineering*, London, 1962; and an impressive account of its application is Thomas P. Hughes, *Networks of Power*, Baltimore and London, 1983. The development of nuclear power is well summarized by Lord Hinton, 'Atomic Energy', in *The History of Technology*, Oxford, Vol. VI, Chapter 10.

1. On the relationship between technology and thermodynamic theory see D. C. L. Cardwell, *From Watt to Clausius – The Rise of Thermodynamics in the Early Industrial Age*, London, 1972.
2. The hot-air engine is the subject of continuing research by Phillips of Eindhoven and others. See G. Walker, *Stirling Engines*, Oxford, 1980. Also D. W. Loveridge, 'Robert Stirling – Preacher and Inventor', in *Transactions of the Newcomen Society*, Vol. 50, 1978–9, pp. 1–10.
3. The political and administrative complexities of nuclear power in Britain are explored by Margaret Gowing in *Britain and Atomic Energy, 1939–1945*, London, 1964, and *Independence and Deterrence – Britain and Atomic Energy, 1945–1952*, 2 vols., London, 1974. There was a flurry of scientific excitement in 1990 regarding the possibility that 'cold' nuclear fusion had been discovered – i.e. without the extremely high temperatures of the hydrogen bomb – but these hopes appear to have been confounded as misinterpretations of experimental 'noise'.

5. The Emergence of the Factory

The Industrial Revolution has been an attractive subject to historians for several generations, so that there is no lack of interesting accounts and interpretations. As a compact factual summary, T. S. Ashton, *The Industrial Revolution 1760–1830*, Oxford, first published 1948, is still difficult to beat. But also worthy of note are R. M. Hartwell, ed., *The Causes of the Industrial Revolution in England*, London, 1967, and the long chapter by David S. Landes, 'Technological Change and Development in Western Europe, 1750–1914' in *The Cambridge Economic History of Europe*, Vol. VI, Cambridge, 1965, subsequently published separately under the title *The Unbound Prometheus*, Cambridge, 1969. P. Mathias, *The First Industrial Nation*, London, 1969, is a very useful general survey.

On agricultural developments, there is a good overview in J. D. Chambers and G. E. Mingay, *The Agricultural Revolution, 1750–1880*, London, 1966. The starting-point for any historical study of mining is the superb illustrated Hoover edition of Georgius Agricola, *De re metallica* (first published in 1556), New York, 1950. On coalmining, J. U. Nef, *The Rise of the British Coal Industry*, London, 2 vols., 1932, is of wider application than just Britain. The same may be said of H. R. Schubert, *History of the British Iron and Steel Industry from c.450 BC to AD 1775*, London, 1957. For the chemical industry, see A. and N. L. Clow, *The Chemical Revolution*, London, 1952, and L. F. Haber, *The Chemical Industry during the Nineteenth Century*, Oxford, 1958. Textile industries have been well covered by historians, although not always with much technical expertise. See, however, W. English, *The Textile Industry*, London, 1969, and Richard L. Hills, *Power in the Industrial Revolution*, Manchester, 1970. Hills has also written *Papermaking in Britain 1488–1988 – A Short History*, London, 1988.

Diane Baker, *Potworks*, published by the Royal Commission on Historical Monuments in England (RCHME), London, 1991, contains a vivid account of this important industry.

The factory itself has received less attention than the processes contained within it. But see J. Tann, *The Development of the Factory*, London, 1970; and, on early factory organization, S. Pollard, *The Genesis of Modern Management*, London, 1963, and Penguin Books, 1968.

1. See Schubert, op. cit., for an account of the direct and indirect processes in iron making.

6. The Age of Mass Production

There is an excellent review of the development of the engineering industry in L. T. C. Rolt, *Tools for the Job: A Short History of Machine Tools*, London, 1965. See also W. H. G. Armytage, *A Social History of Engineering*, London, 1961; and, on the Brunel block-making machinery, K. R. Gilbert, *The Portsmouth Blockmaking Machinery* (Science Museum Monograph), London, 1965. For a recent treatment of the whole theme, see D. Hounshell, *From the American System to Mass-Production*, Baltimore, 1984. H. J. Habakkuk, *American and British Technology in the 19th Century: The Search for Labour-Saving Inventions*, Cambridge, 1962, is an important contribution to our understanding; and S. B. Saul, ed., *Technological Change – The United States and Britain in the 19th Century*, London, 1970, is also useful.

On particular industries in this period, several of the references in the last chapter are appropriate.

Amongst the more specialized recent literature, the following are of interest: Leonard S. Reich, *The Making of American Industrial Research – Science and Business at GE and Bell, 1876–1926*, Cambridge, 1985; Kristine Bruland, *British Technology and European Industrialization*, Cambridge, 1989, gives a perspective from the Norwegian periphery of European industrialization. A classic work is Frederick W. Taylor's *Scientific Management*, 1906. And a good general account of modern industrial organization is still J. A. C. Brown, *The Social Psychology of Industry*, Penguin Books, 1954.

1. L. T. C. Rolt, *Tools for the Job*, summarizes the transmission of engineering expertise between these pioneering figures.
2. There has been a recent article in *Technology and Culture* about the career of Perkin – 'Perkin's Mauve: Ancestor of the Organic Chemical Industry', Vol. 31, No. 1, January 1990, pp. 51–82, by Anthony S. Travis.
3. On Edison, a good introduction is Wyn Wachhorst: *Thomas Alva Edison – An American Myth*, Cambridge, MA, 1981, in which his reputation as 'the Wizard of Menlo Park' is explored.
4. Geoffrey Tweedale, *Sheffield Steel and America: A Century of Commercial and Technological Interdependence 1830–1930*, Cambridge,

1987, pp. 58–65, makes this point in relation to the work of Sir Robert Hadfield.

7. Transport before the Railway Age

Transport history has stimulated much interest and a substantial literature, much of it of a very ephemeral nature. There are some competent chapters in the general histories of technology which give an overview of the developments.

The monumental work by W. T. Jackman, *The Development of Transportation in Modern England*, 1916, 2nd edn, London, 1962, is still a useful work of reference, as is G. Ottley's *A Bibliography of British Railway History*, London, 1965.

Charles Hadfield has written extensively on canals and has edited a series of books on the British canal network. His most general work is *The Canal Age*, Newton Abbot, 1968.

A rather similar role has been performed for academic railway history by Jack Simmons; see, for example, his book *The Railways of Britain*, 1961, 2nd edn, London, 1968. Professor Simmons has recently updated this account in his new work, *The Victorian Railway*, London, 1991.

On lighthouses, see D. B. Hague and R. Christie, *Lighthouses – Their Architecture, History, and Archaeology*, Llandysul, Dyfed, 1975. J. Bird, *The Major Seaports of the United Kingdom*, London, 1963, and Gordon Jackson, *The History and Archaeology of Ports*, Newton Abbot, 1983, give a comparable overview of port development. There is nothing quite as convenient on roads, but see A. Bird, *Roads and Vehicles*, London, 1969.

1. Hugh Malet, *The Canal Duke*, London, 1961, gives a good account of the Duke of Bridgewater, his agents, and Brindley. For a review of these pioneering years of British engineering, see R. A. Buchanan, *The Engineers: A History of the Engineering Profession in Britain 1750–1914*, London, 1989.
2. S. Smiles, *Lives of the Engineers*, London, 1862, reprinted in 3 vols. by David & Charles, with an introduction by L. T. C. Rolt, 1968,

remains a valuable source of information for Brindley, Smeaton, Telford and other early engineers. See also A. W. Skempton, ed., *John Smeaton FRS*, London, 1981.

3. L. T. C. Rolt, *George and Robert Stephenson – The Railway Revolution*, London, 1960, and Penguin Books, 1978, gives a lively account of these events.

4. The story of Brunel and the Great Western Railway is well told in L. T. C. Rolt, *Isambard Kingdom Brunel – A Biography*, London, 1957, and Penguin Books, 1978, new edn, 1989.

8. Transport from Steam Trains to Rockets

The general references carry forward from the previous chapter. On British aspects, a good introduction is Philip S. Bagwell's *The Transport Revolution from 1770*, London, 1974. The development of the railway network over the continents of the world, and the replacement of steam traction by electric and diesel power, have been the subject of much scholarly comment. See, for instance, Michael Duffy, 'Techno-morphology and the Stephenson Traction System', in *Transactions of the Newcomen Society*, Vol. 54, 1982. Note also the lavishly illustrated work by Brian Hollingsworth and Arthur Cook, *The Illustrated Encyclopaedia of the World's Modern Locomotives*, London, 1983.

On the history of the steamship, another well-illustrated book is Peter Kemp's *The History of Ships*, London, 1978. See also the excellent series of ten slim illustrated volumes *The Ship*, published by HMSO for the National Maritime Museum, especially that by A. Preston, *Dreadnought to Nuclear Submarine*, 1980. Similarly, Robert Simper, *Britain's Maritime Heritage*, Newton Abbot, 1982, and D. K. Brown, *A Century of Naval Construction – The History of the Royal Corps of Naval Constructors*, London, 1983, both make valuable contributions to the subject.

Edgar C. Smith, *A Short History of Naval and Marine Engineering*, Cambridge, 1937, is still a valuable summary of the engineering aspects.

For the automobile, there is a mass of specialist literature, covering virtually every manufacturer and type of car. A good general introduction, with breezy text and attractive pictures, is David Hodges and

David B. Wise, *The Story of the Car*, London, 1974. See also J.-P. Bardou, J.-J. Chanaron, P. Fridenson, and J. M. Laux, *La Revolution automobile*, Paris, 1977; R. Hough and L. J. K. Setright, *A History of the World's Motorcycles*, London, 1966; and Eric Tompkins, *The History of the Pneumatic Tyre*, London, 1981. Bicycles are well covered by Andrew Ritchie, *King of the Road – An Illustrated History of Cycling*, London, 1975 – an entertaining popular account.

On the aeroplane and the history of flight, see C. H. Gibbs-Smith, *The Aeroplane – An Historical Survey*, London, 1960. See also R. E. G. Davies, *A History of the World's Airlines*, Oxford, 1964. F. Whittle, *Jet – the Story of a Pioneer*, London, 1953, and paperback, 1957, is a fascinating account of the author's personal involvement. The more recent work by John Golley, *Whittle – The True Story*, Washington, DC, 1987, does not add substantially to Whittle's own version.

1. L. T. C. Rolt, *Isambard Kingdom Brunel – a Biography*, London, 1957, and Penguin Books, 1975, new edn, 1989, presents a vivid account of Brunel's three great ships.
2. See C. K. Harley, 'The Shift from Sailing Ships to Steam Ships, 1850–1890 – a Study in Technological Change and its Diffusion', in D. N. McCloskey, ed., *Essays in a Mature Economy: Britain after 1840*, London, 1971.
3. On the history of the bicycle, see Ritchie, op. cit.
4. Michael J. Neufeld, 'Weimar Culture and Futuristic Technology – The Rocketry and Spaceflight Fad in Germany, 1923–1933', in *Technology and Culture*, Vol. 31, October 1990, No. 4, pp. 725–52.

9. Communications and Information Technology

The *Oxford History of Technology* volumes have useful sections on the development of communications down to 1950. See especially Vol. VII, with D. G. Tucker on 'Electrical communications', Helmut Gernsheim on 'Photography' and David B. Thomas on 'Cinematography'.

Faraday studies have become something of a vogue in recent years: see, for example, L. Pearce Williams, *Michael Faraday – a Biography*, London, 1965. Percy Dunsheath, *A History of Electrical Engineering*,

London, 1962, contains good material on telephony and tele-communications, which can be supplemented by W. A. A. Atherton, *From Compass to Computer: A History of Electrical and Electronics Engineering*, London, 1984. Also of interest are Michael R. Williams, *A History of Computing Technology*, New Jersey, 1985; Ithiel de Sola Pool, ed., *The Social Impact of the Telephone*, Cambridge, MA, 1977; P. R. Morris, *A History of the World Semiconductor Industry*, London, 1990; and Martin Campbell-Kelly, *ICL – a Business and Technical History*, Oxford, 1989 (International Computers (Holdings) Limited was formed by amalgamation in 1968 as the British counterpart to IBM).

1. The Jacquard loom, invented by the Frenchman J. M. Jacquard in 1801, had semi-automatic selection of threads for the complicated patterns used in silk weaving; the process was controlled by means of a series of punched cards mounted above the loom. There is a good account of Charles Babbage and the early development of computer technology in B. V. Bowden, ed., *Faster than Thought*, London, 1953. For a useful summary of Hollerith's career, see M. Campbell-Kelly, op. cit., Chapter 1, and also Arthur L. Norberg, 'High-Technology Calculation in the Early 20th Century: Punched Card Machinery in Business and Government', in *Technology and Culture*, Vol. 31, No. 4, October 1990, pp. 753–79.
2. See Atherton, op. cit., p. 283.

10. Infrastructure – Buildings, Bridges and Services

There is a wealth of architectural literature on the history of buildings and structural techniques. A particularly stimulating work on building materials is Alec Clifton-Taylor, *The Pattern of English Building*, London, 1962. On individual buildings, J. M. Richards, *The Functional Tradition in Early Industrial Buildings*, London, 1958, impressively illustrated by Eric de Mare's photographs.

For other aspects of the infrastructure, see J. P. M. Pannell, *An Illustrated History of Civil Engineering*, London, 1964; H. Shirley Smith, *The World's Greatest Bridges*, London, 1953; Derrick Beckett, *Great*

Buildings of the World – Bridges, London, 1969; Norman Smith, *Man and Water*, London, 1975; G. M. Binnie, *Early Victorian Water Engineers*, London, 1981; Thomas P. Hughes, *Networks of Power – Electrification in Western Society 1880–1930*, Baltimore and London, 1983. Percy Dunsheath, op. cit., (Chapter 9 above), is also useful on electrical supply. There are no comparable overviews for sewage services, but see Denis Smith, 'Sir Joseph William Bazalgette (1819–1891) – engineer to the Metropolitan Board of Works', in *Transactions of the Newcomen Society*, Vol. 58, 1986–7, pp. 89–111. On the gas industry, see T. I. Williams, *A History of the British Gas Industry*, Oxford, 1981, and in the United States, Louis Stotz and Alexander Jamison, *History of the Gas Industry*, New York, 1938.

1. The firm was Samuel Hemming of Clift House, Bristol; see R. A. Buchanan and Neil Cossons, *The Industrial Archaeology of the Bristol Region*, Newton Abbot, 1969, p. 61. For a discussion of the 'Old' and 'New' Iron Age, see Chapter 2 above.
2. Norman Smith, *A History of Dams*, New Jersey, 1972, gives an excellent survey of this important subject.
3. It was the work of an Irish engineer, C. Y. O'Connor, who was appointed engineer in chief to Western Australia in 1891; see R. A. Buchanan, *The Engineers: A History of the Engineering Profession in Britain 1750–1914*, London, 1989, p. 155.

11. Technology and People

The social context of technological revolution is a many-sided subject and there is as yet no satisfactory overview of it. But there are many useful studies of particular themes, some of which are noted below. For reflections on demographic change, see E. A. Wrigley, *Continuity, Chance and Change*, Cambridge, 1988. Brooke Hindle provides a stimulating insight into engineering creativity in *Emulation and Invention*, New York, 1981; a related dimension of creative experience is explored, with fascinating illustrative material, by Francis D. Klingender: *Art and the Industrial Revolution*, London, 1947, revised edition by Arthur Elton, London, 1968. Arnold Pacey, *Technology in World Civilization*,

Oxford, 1990, is generally aware of the issues, as he was, indeed, in his earlier works, *The Maze of Ingenuity*, London, 1974, and *The Culture of Technology*, Oxford, 1983.

1. Wyn Wachhorst, *Thomas Alva Edison – an American Myth*, Cambridge, MA, 1981, pp. 13–14.
2. J. Jewkes, D. Sawers and R. Stillerman, *The Sources of Invention*, 2nd edn, London, 1969.
3. For the development of patent law, see Christine Macleod, *Inventing the Industrial Revolution: The English Patent System, 1600–1800*, Cambridge, 1988.
4. The thesis of population increase and the 'ecological niche' has been forcefully expressed by Richard G. Wilkinson, *Poverty and Progress: An Ecological Model of Economic Development*, London, 1973.
5. Medieval diet figures prominently in Lynn White Jr, *Medieval Technology and Social Change*, Oxford, 1960.
6. Asa Briggs, *Victorian Cities*, London, 1963, p. 15, quotes Richard Hoggart as the author of this expression. On urban transport, an excellent study is that by T. C. Barker and M. Robbins, *The History of London Transport: Passenger Travel and the Development of the Metropolis*, London, 2 vols., 1963 and 1974.
7. Ebenezer Howard, *Garden Cities of To-morrow*, London, 1902, pioneered the idea of planned cities for the modern world. The subject has generated much literature, of which an excellent recent example has been Peter Hall, *Cities of Tomorrow*, Oxford, 1988.

12. Technology and the State

Some of the implications of technology for the political life of the state, nationalism, and related concepts, are explored in R. A. Buchanan, *History and Industrial Civilisation*, London, 1979. See also Eric J. Hobsbawm, *The Age of Revolution – Europe 1789–1848*, London, 1962, and *Nations and Nationalism since 1780*, Cambridge, 1990. On the professionalization and training of engineers, see R. A. Buchanan, *The Engineers: A History of the Engineering Profession in Britain 1750–1914*, London, 1989. Also D. S. L. Cardwell, *The Organisation of Science in*

England, London, 1957, and M. Sanderson, *The Universities and British Industry 1850–1970*, London, 1972. Thomas S. Kuhn, *The Structure of Scientific Revolutions*, Chicago, 1962, has been a seminal work in our understanding of the social relations of science. The recent work on military history, Geoffrey Parker, *The Military Revolution: Military Innovation and the Rise of the West, 1500–1800*, Cambridge, 1988, is stimulating but deals only with the early part of our period. J. U. Nef, *War and Human Progress*, New York, 1963, is a sustained critique of the view that war is a benefit to civilization.

1. These famous words from Thomas Hobbes, *The Leviathan*, 1651, come in a passage in Chapter 13 of Part I, where he is discussing the results of the loss of security which come with war and the breakdown of society.
2. The development of the English patent system is well described by Christine MacLeod in *Inventing the Industrial Revolution: The English Patent System, 1660–1800*, Cambridge, 1988. S. B. Saul, 'The nature and diffusion of technology', in A. J. Youngson, ed., *Economic Development in the Long Run*, London, 1972, has perceptive comments on the effects of patents: 'The sale of patent rights became a major form of technological diffusion by the end of the nineteenth century' (p. 56).
3. Daniel R. Headrick has written illuminatingly on this subject: see his article 'The Tools of Imperialism: Technology and the Expansion of European Colonial Empires in the Nineteenth Century', in *Journal of Modern History*, June 1979, subsequently expanded into *The Tools of Empire: Technology and European Imperialism in the Nineteenth Century*, New York and Oxford, 1981. Carlo Cipolla, *Guns and Sails*, London, 1965, is also interesting.
4. For the background to the foundation of the Royal Society, see Charles Webster, *The Great Instauration: Science, Medicine and Reform 1626–1660*, London, 1974 – following Bacon's philosophical programme, his *Instauratio magna*, 'the puritan intellectuals became committed to a dedicated attempt to procure the return of man's dominion over nature' (p. xvi). Benjamin Farrington, *Francis Bacon – Philosopher of Industrial Science*, London, 1951, is a useful summary of Bacon's work. And an interesting sociological view is provided by

Robert K. Merton, *Science, Technology and Society in Seventeenth Century England*, New York, 1970, and Sussex, 1978.
5. W. H. G. Armytage: *A Social History of Engineering*, London, 1961: 'The first world war was a chemists' war'. p. 251. The extension of the idea to describe the Second World War as 'a physicists' war' was suggested by Hilary and Steven Rose: *Science and Society*, Penguin 1969. See also William H. McNeill: *The Pursuit of Power – Technology, Armed Force and Society since AD 1000*, Chicago, 1982.

13. The Technological Dilemma

This chapter covers a wide range of environmentalist and ecological issues on which much has been written in recent decades. These subjects have generated considerable agitation and controversy and have led to some significant political action such as the British clean-air legislation. I do not know of any single account which covers the whole range satisfactorily, but there have been a handful of stimulating works, several of which are given in the following notes.

1. Derek J. de Solla Price, *Little Science, Big Science*, New York, 1963: it is a slim book, consisting of the four George B. Pegram Lectures for 1962.
2. Rachel Carson, *Silent Spring*, Boston, Mass., 1962, London, 1963, and Penguin Books, 1965.
3. James Lovelock, *Gaia – A New Look at Life on Earth*, Oxford, 1979.
4. Jacques Ellul, *The Technological Society*, London, 1965, p. 411.
5. George Orwell, *Nineteen Eighty-four*, 1949, and Penguin Books, 1954: 'If you want a picture of the future, imagine a boot stamping on a human face – for ever' (p. 215 of the Penguin edition).
6. Thomas P. Hughes, *Networks of Power – Electrification in Western Society 1880–1930*, Baltimore and London, 1983, develops the concept in relation to systems of electric power.
7. David Collingridge, *The Social Control of Technology*, London, 1980, pp. 13–15.
8. E. Schumacher, *Small is Beautiful: A Study of Economics as if People*

Mattered, London, 1973, and paperback edition, 1974.

9. The Club of Rome was responsible for the report by Donella H. Meadows *et al.*, *The Limits to Growth*, London, 1972. It elicited a prompt riposte from H. S. D. Cole *et al.*, *Thinking about the Future*, London, 1973.

10. The Report of the Brandt Commission was published in 1980 as *North–South: A Programme for Survival*.

14. The Way to the Stars

There is a good discussion of progress by Sidney Pollard, *The Idea of Progress*, London, 1968, and Penguin Books, 1971. A pioneering work on the scope of space exploration, before the first satellite had been launched, was Arthur C. Clarke, *The Exploration of Space*, London, 1951, published in a revised edition by Penguin Books, 1958. Clarke is one of the outstanding modern writers of science fiction, evoking brilliant and haunting images of what could be achieved by space exploration. But he also contributed to the conception of satellite communication systems with his idea, published in 1945, for using satellites placed in stationary orbits to relay messages instantly to all points on the surface of the Earth. The actual achievements of space exploration have spawned an enormous literature, much of it of a very transitory quality. But for an up-to-date, if rather breathless, review of accomplishments, see Thomas R. McDonough, *Space – The Next Twenty-five Years*, New York, 1987, revised and updated 1989. *The Cambridge Atlas of Astronomy*, ed. Jean Audouze and Guy Israel, Cambridge, 2nd edn, 1988, is a constant source of inspiration.

1. *The Communist Manifesto* of 1848 is redolent with an idea of progress. Although Charles Darwin, in *The Origin of Species*, 1859, and *The Descent of Man*, 1871, expounded a carefully elaborated theory of evolution by natural selection, he was cautious about any notion of moral progress and only adopted the idea of the survival of the fittest from Herbert Spencer, who had a much more positively progressivistic attitude.

2. Clarke's paper of 1945, 'The Space-Station: Its Radio Applications', is reprinted in Arthur C. Clarke, *Ascent to Orbit – A Scientific Autobiography*, New York, 1984. A more generally prophetic work by Clarke is his *Profiles of the Future*, London, 1962 and paperback, 1964, in which he explores the possibilities of future developments in human investigations.

Index

Index

tin, 11, 90–91, 107
Titanic, SS, 162
toll-making
  industrial, 104–5
  origins of, 8–10
town planning, 212–13
traffic engineering, 213
training, *see* education
trams, 211, 212
transistors, 166–7
transport
  construction and, 178–9
  electricity and, 74, 139, 140–41, 151, 211
  food, 116
  urban and suburban, 210–11, 212
  see also individual systems
Trent and Mersey Canal, 126
Trevithick, Richard, 56–7, 133, 186
Triewald, Marten, 50
Tsiolkovsky, Konstantin, 259
tunnels, 184–6
turbine(s)
  gas, 70, 77, 154–5, 156, 234
  steam, 55, 58, 61–4, 76, 77, 139, 145, 192
  water, 59, 61, 192–3
Turbinia, 64, 145
Turing, Alan M., 166
Turner, J.M.W., 59
turnpike trusts, 129–30
typewriters, 175, 200

ultraviolet light, 207
underground railways, 140, 185
United States of America
  aeroplane industry in, 153, 155
  agriculture in, 86, 104, 116
  automobile industry in, 68, 105, 117–18, 149, 150

bridges in, 182
buildings in, 181
coal industry in, 89
communications in, 159, 161–2, 163, 164, 165–7, 172
electricity supply in, 73
managerialism in, 117–18, 120
mass production in, 68, 104, 105, 116, 117–18, 150
oil industry in, 110–11
railways in, 135, 140
space exploration by, 72, 259–67
spread of inventions to, 27, 30–31
steam engines in, 50, 56
steel industry in, 106–7
warfare and, 234–5
Uranus, exploration of, 266
urbanization, 209–13

V1 and V2 weapons, 72, 156, 234, 259–60
vaccination, 206–7
Venus, exploration of, 262
Verne, Jules, 259
viaducts, 184
Volta, Alessandro, 73
Voyager space probes, 266–7

Wade, General, 130
Wankel engine, 71, 150
warfare, 231–5, 238–9, 249–50
waste
  domestic, 190–91, 206, 210, 211
  radioactive, 78, 239
water power, 45–6, 47, 59, 61, 79–80, 192–3
  textiles industry and, 96, 99
water supply, 186–91, 205–6, 210, 211
water turbine, 59, 61, 192–3
water-wheels, 45–6, 47, 59

298